Linear Mathematics and Its Applications

Larry J. Goldstein, David C. Lay, and David I. Schneider

— Matrices

— Linear Programming, A Geometric Approach

— The Simplex Method

Excerpted from:
Mathematics for the Management, Life, and Social Sciences
Second Edition
by Goldstein, Lay, and Schneider

Chapters 1, 2, and 3 of this book were previously published as chapters 5, 6, and 7 of *Mathematics for the Management, Life, and Social Sciences,* Second Edition (1984). Prentice-Hall, Inc., Englewood Cliffs, New Jersey 07632.

BA 2565

ISBN 0–536–05726–5

Printed in the United States of America

GINN PRESS

160 Gould Street/Needham Heights, MA 02194
Simon & Schuster Higher Education Publishing Group

Contents

Chapter 1

Matrices

We begin this chapter by developing a method for solving systems of linear equations in any number of variables. Our discussion of this method will lead naturally into the study of mathematical objects called *matrices*. The arithmetic and applications of matrices are the main topics of the chapter. We will discuss in detail the application of matrix arithmetic to input-output analysis, which can be (and is) used to make production decisions for large businesses and entire economies.

1.1 Solving Systems of Linear Equations, I

In an earlier chapter we presented a method for solving systems of linear equations in two variables. The method of Chapter 2* is very efficient for determining the solutions. Unfortunately, it works only for systems of linear equations having *two* variables. In many applications we meet systems having more than two variables, as the following example illustrates.

EXAMPLE 1 A bacteria culture contains three species of bacteria. Each species requires certain amounts of three different nutrients: a nitrogen source, a carbon source, and a phosphate source. The daily requirements (expressed in appropriate units) of each nutrient are summarized in the accompanying chart.

	Nutrient		
Species	*Nitrogen source*	*Carbon source*	*Phosphate source*
A	1 unit/day	3 units/day	4 units/day
B	2 units/day	5 units/day	7 units/day
C	2 units/day	5 units/day	8 units/day

* The reader is referred to ***Mathematics for the Management, Life, and Social Sciences* by**
Goldstein, Lay, and Schneider

Every day the culture is supplied with 15,000 units of a nitrogen source, 40,000 units of a carbon source, and 59,000 units of a phosphate source. Assume that all the nutrients are utilized. How many of each species of bacteria can the culture simultaneously support?

Solution Let us express the given data by a system of equations. Let x be the number of bacteria of species A, y the number of bacteria of species B, and z the number of bacteria of species C. The first piece of data is that there are 15,000 units per day of the nitrogen source. Therefore, we must have

$$\begin{aligned}&[\text{amount of nitrogen source for species A}]\\ &\qquad + [\text{amount of nitrogen source for species B}]\\ &\qquad + [\text{amount of nitrogen source for species C}] = 15{,}000.\end{aligned}$$

Thus, from the first column of the chart, we have

$$\begin{aligned}&1 \cdot [\text{number of species A bacteria}] + 2 \cdot [\text{number of species B bacteria}]\\ &\qquad + 2 \cdot [\text{number of species C bacteria}] = 15{,}000.\end{aligned}$$

That is,

$$x + 2y + 2z = 15{,}000.$$

In a similar way, we derive equations for carbon and phosphate:

$$3x + 5y + 5z = 40{,}000$$

$$4x + 7y + 8z = 59{,}000.$$

So we see that the numbers x, y, z of the three species in the culture must simultaneously satisfy these linear equations in three variables:

$$\begin{cases} x + 2y + 2z = 15{,}000 \\ 3x + 5y + 5z = 40{,}000 \\ 4x + 7y + 8z = 59{,}000. \end{cases} \tag{1}$$

Later we shall give a method for determining the solution of this system. It will be easy to show that the only solution is $x = 5000$, $y = 1000$, $z = 4000$. Once we are given the solution, it is easy to check that these values of x, y, z make all three equations true:

$$5000 + 2 \cdot 1000 + 2 \cdot 4000 = 15{,}000$$

$$3 \cdot 5000 + 5 \cdot 1000 + 5 \cdot 4000 = 40{,}000$$

$$4 \cdot 5000 + 7 \cdot 1000 + 8 \cdot 4000 = 59{,}000.$$

Thus the culture can support 5000 bacteria of species A, 1000 bacteria of species B, and 4000 bacteria of species C.

In this section we develop a step-by-step procedure for solving systems of linear equations such as (1). The procedure, called the *Gaussian elimination method,* consists of repeatedly simplifying the system, using so-called elementary row operations, until the solution stares us in the face!

In the system of linear equations (1) the equations have been written in such a way that the x-terms, the y-terms, and the z-terms lie in different columns. We shall always be careful to display systems of equations with separate columns for each variable. One of the key ideas of the Gaussian elimination method is to think of the solution as a system of linear equations in its own right. For example, we can write the solution of the system (1) as:

$$\begin{cases} x = 5000 \\ y = 1000 \\ z = 4000. \end{cases} \tag{2}$$

This is just a system of linear equations in which the coefficients of most terms are zero! Since the only terms with nonzero coefficients are arranged on a diagonal, such a system is said to be in *diagonal form.*

Our method for solving a system of linear equations consists of repeatedly using three operations which alter the system but do not change the solutions. The operations are used to transform the system into a system in diagonal form. Since the operations involve only elementary arithmetic and are applied to entire equations (i.e., rows of the system), they are called *elementary row operations.* Let us begin our study of the Gaussian elimination method by introducing these operations.

Elementary Row Operation 1 Rearrange the equations in any order.

This operation is harmless enough. It certainly does not change the solutions of the system.

Elementary Row Operation 2 Multiply an equation by a nonzero number.

For example, if we are given the system of linear equations

$$\begin{cases} 2x - 3y + 4z = 11 \\ 4x - 19y + z = 31 \\ 5x + 7y - z = 12, \end{cases}$$

then we may replace it by a new system obtained by leaving the last two equations unchanged and multiplying the first equation by 3. To accomplish this, multiply each term of the first equation by 3. The transformed system is

$$\begin{cases} 6x - 9y + 12z = 33 \\ 4x - 19y + z = 31 \\ 5x + 7y - z = 12. \end{cases}$$

The operation of multiplying an equation by a nonzero number does not change the solutions of the system. For if a particular set of values of the variables satisfy the original equation, they satisfy the resulting equation, and vice versa.

Elementary row operation 2 may be used to make the coefficient of a particular variable 1.

EXAMPLE 2 Replace the system

$$\begin{cases} -5x + 10y + 20z = 4 \\ x - 12z = 1 \\ x + y + z = 0 \end{cases}$$

by an equivalent system in which the coefficient of x in the first equation is 1.

Solution The coefficient of x in the first equation is -5, so we use elementary row operation 2 to multiply the first equation by $-\frac{1}{5}$. Multiplying each term of the first equation by $-\frac{1}{5}$ gives

$$\begin{cases} x - 2y - 4z = -\frac{4}{5} \\ x - 12z = 1 \\ x + y + z = 0. \end{cases}$$

Another operation that can be performed on a system without changing its solutions is to replace one equation by its sum with some other equation. For example, consider this system of equations:

$$A: \begin{cases} x + y - 2z = 3 \\ x + 2y - 5z = 4 \\ 5x + 8y - 18z = 14. \end{cases}$$

We can replace the second equation by the sum of the first and the second. Since

$$\begin{array}{r} x + y - 2z = 3 \\ + \quad x + 2y - 5z = 4 \\ \hline 2x + 3y - 7z = 7, \end{array}$$

the resulting system is

$$B: \begin{cases} x + y - 2z = 3 \\ 2x + 3y - 7z = 7 \\ 5x + 8y - 18z = 14. \end{cases}$$

If a particular choice of x, y, z satisfies system A, it also satisfies system B. This is because system B results from adding equations. Similarly, system A can be derived from system B by subtracting equations. So any particular solution of system A is a solution of system B, and vice versa.

The operation of adding equations is usually used in conjunction with elementary row operation 2. That is, an equation is changed by adding to it a nonzero multiple of another equation. For example, consider the system

$$\begin{cases} x + y - 2z = 3 \\ x + 2y - 5z = 4 \\ 5x + 8y - 18z = 14. \end{cases}$$

Let us change the second equation by adding to it twice the first. Since

$$\begin{array}{rr} \text{2(first)} & 2x + 2y - 4z = 6 \\ + \text{(second)} & \underline{x + 2y - 5z = 4} \\ & 3x + 4y - 9z = 10, \end{array}$$

the new second equation is

$$3x + 4y - 9z = 10$$

and the transformed system is

$$\begin{cases} x + y - 2z = 3 \\ 3x + 4y - 9z = 10 \\ 5x + 8y - 18z = 14. \end{cases}$$

Since addition of equations and elementary row operation 2 are often used together, let us define a third elementary row operation to be the combination:

Elementary Row Operation 3 Change an equation by adding to it a multiple of another equation.

For reference, let us summarize the elementary row operations we have just defined.

Elementary Row Operations

1. Rearrange the equations in any order.
2. Multiply an equation by a nonzero number.
3. Change an equation by adding to it a multiple of another equation.

The idea of the Gaussian elimination method is to transform an arbitrary system of linear equations into diagonal form by repeated application of the three elementary row operations. To see how the method works, consider the following example.

EXAMPLE 3 Solve the following system by the Gaussian elimination method.

$$\begin{cases} x - 3y = 7 \\ -3x + 4y = -1. \end{cases}$$

Solution Let us transform this system into diagonal form by examining one column at a time, starting from the left. Examine the first column:

$$\boxed{\begin{array}{r} x \\ -3x \end{array}}$$

The coefficient of the top x is 1, which is exactly what it should be for the system to be in diagonal form. So we do nothing to this term. Now examine the next term in the column, $-3x$. In diagonal form this term must be absent. In order to accomplish this we add a multiple of the first equation to the second. Since the coefficient of x in the second is -3, we add three times the first equation to the second equation in order to cancel the x-term. (Abbreviation: $[2] + 3[1]$. The $[2]$ means that we are changing equation 2. The expression $[2] + 3[1]$ means that we are replacing equation 2 by the original equation plus three times equation 1.)

$$\begin{cases} x - 3y = 7 \\ -3x + 4y = -1 \end{cases} \xrightarrow{[2] + 3[1]} \begin{cases} x - 3y = 7 \\ \quad - 5y = 20. \end{cases}$$

The first column now has the proper form, so we proceed to the second column. In diagonal form that column will have one nonzero term, namely the second, and the coefficient of y in that term must be 1. To bring this about, multiply the second equation by $-\frac{1}{5}$ (abbreviation: $-\frac{1}{5}[2]$):

$$\begin{cases} x - 3y = 7 \\ \quad - 5y = 20 \end{cases} \xrightarrow{-\frac{1}{5}[2]} \begin{cases} x - 3y = 7 \\ \qquad y = -4. \end{cases}$$

The second column still does not have the correct form. We must get rid of the $-3y$ term in the first equation. We do this by adding a multiple of the second equation to the first. Since the coefficient of the term to be canceled is -3, we add three times the second equation to the first:

$$\begin{cases} x - 3y = 7 \\ \qquad y = -4 \end{cases} \xrightarrow{[1] + 3[2]} \begin{cases} x \qquad = -5 \\ \qquad y = -4. \end{cases}$$

The system is now in diagonal form and the solution can be read off: $x = -5$, $y = -4$.

EXAMPLE 4 Use the Gaussian elimination method to solve the system

$$\begin{cases} 2x - 6y = -8 \\ -5x + 13y = 1. \end{cases}$$

Solution We can perform the calculations in a mechanical way, proceeding column by column from the left:

$$\begin{cases} 2x - 6y = -8 \\ -5x + 13y = 1 \end{cases} \xrightarrow{\frac{1}{2}[1]} \begin{cases} x - 3y = -4 \\ -5x + 13y = 1 \end{cases}$$

$$\xrightarrow{[2] + 5(1)} \begin{cases} x - 3y = -4 \\ -2y = -19 \end{cases}$$

$$\xrightarrow{-\frac{1}{2}[2]} \begin{cases} x - 3y = -4 \\ y = \frac{19}{2} \end{cases}$$

$$\xrightarrow{[1] + 3[2]} \begin{cases} x = \frac{49}{2} \\ y = \frac{19}{2}. \end{cases}$$

So the solution of the system is $x = \frac{49}{2}$, $y = \frac{19}{2}$.

The calculation becomes easier to follow if we omit writing down the variables at each stage and work only with the coefficients. At each stage of the computation the system is represented by a rectangular array of numbers. For instance, the original system is written*

$$\left[\begin{array}{cc|c} 2 & -6 & -8 \\ -5 & 13 & 1 \end{array}\right].$$

The elementary row operations are performed on the rows of this rectangular array just as if the variables were there. So, for example, the first step above is to multiply the first equation by $\frac{1}{2}$. This corresponds to multiplying the first row of the array by $\frac{1}{2}$ to get

$$\left[\begin{array}{cc|c} 1 & -3 & -4 \\ -5 & 13 & 1 \end{array}\right].$$

The diagonal form just corresponds to the array

$$\left[\begin{array}{cc|c} 1 & 0 & \frac{49}{2} \\ 0 & 1 & \frac{19}{2} \end{array}\right].$$

Note that this array has ones down the diagonal and zeros everywhere else on the left. The solution of the system appears on the right.

A rectangular array of numbers is called a *matrix* (plural: *matrices*). In the next example, we use matrices to carry out the Gaussian elimination method.

* The vertical line between the second and third columns is a placemarker which separates the data obtained from the right- and left-hand sides of the equations. It is inserted for visual convenience.

EXAMPLE 5 Use the Gaussian elimination method to solve the system

$$\begin{cases} 3x - 6y + 9z = 0 \\ 4x - 6y + 8z = -4 \\ -2x - y + z = 7. \end{cases}$$

Solution The initial array corresponding to the system is

$$\left[\begin{array}{ccc|c} 3 & -6 & 9 & 0 \\ 4 & -6 & 8 & -4 \\ -2 & -1 & 1 & 7 \end{array}\right].$$

We must use elementary row operations to transform this array into diagonal form—that is, with ones down the diagonal and zeros everywhere else on the left:

$$\left[\begin{array}{ccc|c} 1 & 0 & 0 & * \\ 0 & 1 & 0 & * \\ 0 & 0 & 1 & * \end{array}\right].$$

We proceed one column at a time.

$$\left[\begin{array}{ccc|c} 3 & -6 & 9 & 0 \\ 4 & -6 & 8 & -4 \\ -2 & -1 & 1 & 7 \end{array}\right] \xrightarrow{\frac{1}{3}[1]} \left[\begin{array}{ccc|c} 1 & -2 & 3 & 0 \\ 4 & -6 & 8 & -4 \\ -2 & -1 & 1 & 7 \end{array}\right] \xrightarrow{[2] + (-4)[1]}$$

$$\left[\begin{array}{ccc|c} 1 & -2 & 3 & 0 \\ 0 & 2 & -4 & -4 \\ -2 & -1 & 1 & 7 \end{array}\right] \xrightarrow{[3] + 2[1]} \left[\begin{array}{ccc|c} 1 & -2 & 3 & 0 \\ 0 & 2 & -4 & -4 \\ 0 & -5 & 7 & 7 \end{array}\right] \xrightarrow{\frac{1}{2}[2]}$$

$$\left[\begin{array}{ccc|c} 1 & -2 & 3 & 0 \\ 0 & 1 & -2 & -2 \\ 0 & -5 & 7 & 7 \end{array}\right] \xrightarrow{[1] + 2[2]} \left[\begin{array}{ccc|c} 1 & 0 & -1 & -4 \\ 0 & 1 & -2 & -2 \\ 0 & -5 & 7 & 7 \end{array}\right] \xrightarrow{[3] + 5[2]}$$

$$\left[\begin{array}{ccc|c} 1 & 0 & -1 & -4 \\ 0 & 1 & -2 & -2 \\ 0 & 0 & -3 & -3 \end{array}\right] \xrightarrow{(-\frac{1}{3})[3]} \left[\begin{array}{ccc|c} 1 & 0 & -1 & -4 \\ 0 & 1 & -2 & -2 \\ 0 & 0 & 1 & 1 \end{array}\right] \xrightarrow{[1] + 1[3]}$$

$$\left[\begin{array}{ccc|c} 1 & 0 & 0 & -3 \\ 0 & 1 & -2 & -2 \\ 0 & 0 & 1 & 1 \end{array}\right] \xrightarrow{[2] + 2[3]} \left[\begin{array}{ccc|c} 1 & 0 & 0 & -3 \\ 0 & 1 & 0 & 0 \\ 0 & 0 & 1 & 1 \end{array}\right].$$

The last array is in diagonal form, so we just put back the variables and read off the solution:

$$x = -3, \qquad y = 0, \qquad z = 1.$$

Because so much arithmetic has been performed, it is a good idea to check the solution by substituting the values for x, y, z into each of the equations of the original system. This will uncover any arithmetic errors that may have occurred.

$$\begin{cases} 3x - 6y + 9z = 0 \\ 4x - 6y + 8z = -4 \\ -2x - y + z = 7 \end{cases} \qquad \begin{cases} 3(-3) - 6(0) + 9(1) = 0 \\ 4(-3) - 6(0) + 8(1) = -4 \\ -2(-3) - (0) + (1) = 7 \end{cases}$$

$$\begin{cases} -9 - 0 + 9 = 0 \\ -12 - 0 + 8 = -4 \\ 6 - 0 + 1 = 7 \end{cases}$$

$$\begin{cases} 0 = 0 \\ -4 = -4 \\ 7 = 7. \end{cases}$$

So we have indeed found a solution of the system.

Remark Note that so far we have not had to use elementary row operation 1, which allows interchange of equations. But in some examples it is definitely needed. Consider this system:

$$\begin{cases} \phantom{3x -{}} y + z = 0 \\ 3x - y + z = 6 \\ 6x \phantom{{}- y} - z = 3. \end{cases}$$

The first step of the Gaussian elimination method consists of making the x-coefficient 1 in the first equation. But we cannot do this, since the first equation does not involve x. To remedy this difficulty, just interchange the first two equations to guarantee that the first equation involves x. Now proceed as before. Of course, in terms of the matrix of coefficients, interchanging equations corresponds to interchanging rows of the matrix.

PRACTICE PROBLEMS 1

1. Determine whether the following systems of linear equations are in diagonal form.

(a) $\begin{cases} x \phantom{{}+y} + z = 3 \\ \phantom{x+{}} y \phantom{{}+z} = 2 \\ \phantom{x+y+{}} z = 7 \end{cases}$ (b) $\begin{cases} x \phantom{{}+y+z} = 3 \\ \phantom{x+{}} y = 5 \\ z \phantom{{}+y} = 7 \end{cases}$ (c) $\begin{cases} x \phantom{{}+y+3z} = -1 \\ \phantom{x+{}} y \phantom{{}+3z} = 0 \\ \phantom{x+y+{}} 3z = 4 \end{cases}$

2. Perform the indicated elementary row operation.

(a) $\begin{cases} x - 3y = 2 \\ 2x + 3y = 5 \end{cases} \xrightarrow{[2] + (-2)[1]}$ (b) $\begin{cases} x + y = 3 \\ -x + 2y = 5 \end{cases} \xrightarrow{[2] + (1)[1]}$

3. State the next elementary row operation which should be performed when applying the Gaussian elimination method.

(a) $\left[\begin{array}{rrr|r} 0 & 2 & 4 & 1 \\ 0 & 3 & -7 & 0 \\ 3 & 6 & -3 & 3 \end{array}\right]$ (b) $\left[\begin{array}{rrr|r} 1 & -3 & 4 & 5 \\ 0 & 2 & 3 & 4 \\ -6 & 5 & -7 & 0 \end{array}\right]$

EXERCISES 1

In Exercises 1–8, perform the indicated elementary row operations and give their abbreviations.

1. Operation 2: multiply the first equation by 2.

$$\begin{cases} \frac{1}{2}x - 3y = 2 \\ 5x + 4y = 1. \end{cases}$$

2. Operation 2: multiply the second equation by -1.

$$\begin{cases} x + 4y = 6 \\ \quad\ -y = 2. \end{cases}$$

3. Operation 3: change the second equation by adding to it 5 times the first equation.

$$\begin{cases} \quad x + 2y = 3 \\ -5x + 4y = 1. \end{cases}$$

4. Operation 3: change the second equation by adding to it $(-\frac{1}{2})$ times the first equation.

$$\begin{cases} \ x - 6y = 4 \\ \frac{1}{2}x + 2y = 1. \end{cases}$$

5. Operation 3: change the third equation by adding to it (-4) times the first equation.

$$\begin{cases} \ \ x - 2y + \ z = 0 \\ \qquad y - 2z = 4 \\ 4x + \ y + 3z = 5. \end{cases}$$

6. Operation 3: change the third equation by adding to it 3 times the second equation.

$$\begin{cases} x + 6y - 4z = 1 \\ \qquad y + 3z = 1 \\ \quad -3y + 7z = 2. \end{cases}$$

7. Operation 3: change the first row by adding to it $\frac{1}{2}$ times the second row.

$$\left[\begin{array}{cc|c} 1 & -\frac{1}{2} & 3 \\ 0 & 1 & 4 \end{array}\right].$$

8. Operation 3: change the third row by adding to it (-4) times the second row.

$$\left[\begin{array}{ccc|c} 1 & 0 & 7 & 9 \\ 0 & 1 & -2 & 3 \\ 0 & 4 & 8 & 5 \end{array}\right].$$

In Exercises 9–16, state the next elementary row operation which must be performed in order to put the matrix into diagonal form. Do not perform the operation.

9. $\left[\begin{array}{cc|c} 1 & -5 & 1 \\ -2 & 4 & 6 \end{array}\right]$

10. $\left[\begin{array}{cc|c} 1 & 3 & 4 \\ 0 & 2 & 6 \end{array}\right]$

11. $\left[\begin{array}{cc|c} 1 & 2 & 3 \\ 0 & 1 & 4 \end{array}\right]$

12. $\left[\begin{array}{rrr|r} 1 & -2 & 5 & 7 \\ 0 & -3 & 6 & 9 \\ 4 & 5 & -6 & 7 \end{array}\right]$

13. $\left[\begin{array}{rrr|r} 0 & 5 & -3 & 6 \\ 2 & -3 & 4 & 5 \\ 4 & 1 & -7 & 8 \end{array}\right]$

14. $\left[\begin{array}{rrr|r} 1 & 4 & -2 & 5 \\ 0 & -3 & 6 & 9 \\ 0 & 4 & 3 & 1 \end{array}\right]$

15. $\left[\begin{array}{rrr|r} 1 & 0 & 3 & 4 \\ 0 & 1 & 2 & 5 \\ 0 & 0 & 1 & 6 \end{array}\right]$

16. $\left[\begin{array}{rrr|r} 1 & 2 & 4 & 5 \\ 0 & 0 & 3 & 6 \\ 0 & 1 & 1 & 7 \end{array}\right]$

Solve the following linear systems by using the Gaussian elimination method.

17. $\begin{cases} 3x + 9y = 6 \\ 2x + 8y = 6 \end{cases}$

18. $\begin{cases} \frac{1}{3}x + 2y = 1 \\ -2x - 4y = 6 \end{cases}$

19. $\begin{cases} x - 3y + 4z = 1 \\ 4x - 10y + 10z = 4 \\ -3x + 9y - 5z = -6 \end{cases}$

20. $\begin{cases} \frac{1}{2}x + y = 4 \\ -4x - 7y + 3z = -31 \\ 6x + 14y + 7z = 50 \end{cases}$

21. $\begin{cases} 2x - 2y = -4 \\ 3x + 4y = 1 \end{cases}$

22. $\begin{cases} 2x + 3y = 4 \\ -x + 2y = -2 \end{cases}$

23. $\begin{cases} 4x - 4y + 4z = -8 \\ x - 2y - 2z = -1 \\ 2x + y + 3z = 1 \end{cases}$

24. $\begin{cases} x + 2y + 2z = 11 \\ x - y - z = -4 \\ 2x + 5y + 9z = 39 \end{cases}$

25. A bank wishes to invest a \$100,000 trust fund in three sources—bonds paying 8%, certificates of deposit paying 7%, and first mortgages paying 10%. The bank wishes to realize an \$8000 annual income from the investment. A condition of the trust is that the total amount invested in bonds and certificates of deposit must be triple the amount invested in mortgages. How much should the bank invest in each possible category? Let x, y, and z, respectively, be the amounts invested in bonds, certificates of deposit, and first mortgages. Solve the system of equations by the Gaussian elimination method.

26. A dietician wishes to plan a meal around three foods. Each ounce of food I contains 10% of the daily requirements for carbohydrates, 10% for protein, and 15% for vitamin C. Each ounce of food II contains 10% of the daily requirements for carbohydrates, 5% for protein, and 0% for vitamin C. Each ounce of food III contains 10% of the daily requirements for carbohydrates, 25% for protein, and 10% for vitamin C. How many ounces of each food should be served in order to supply exactly the daily requirements for each of carbohydrates, protein and vitamin C? Let x, y, and z, respectively, be the number of ounces of foods I, II, and III.

SOLUTIONS TO PRACTICE PROBLEMS 1

1. (a) Not in diagonal form, since the first equation contains both x and z.

   (b) Not in diagonal form, since the variables are not arranged in diagonal fashion.

   (c) Not in diagonal form, since the coefficient of z is not 1.

2. (a) Change the system into another system in which the second equation is altered by having (-2) (first equation) added to it. The new system is

$$\begin{cases} x - 3y = 2 \\ \quad\;\; 9y = 1. \end{cases}$$

The equation $9y = 1$ was obtained as follows:

$$\begin{array}{lr} (-2)(\text{first equation}) & -2x + 6y = -4 \\ +\ (\text{second equation}) & 2x + 3y = 5 \\ \hline & 9y = 1. \end{array}$$

(b) Replace the second equation by the first equation multiplied by 1 and added to the second. This is the same as adding the first equation to the second. The result is

$$\begin{cases} x + \;\; y = 3 \\ \quad\;\; 3y = 8. \end{cases}$$

3. (a) The first row should contain a nonzero number as its first entry. This can be accomplished by interchanging the first and third rows.

(b) The first column can be put into proper form by eliminating the -6. To accomplish this, multiply the first row by 6 and add this product to the third row. The notation for this operation is

$$\xrightarrow{[3] + 6[1]}$$

1.2 Solving Systems of Linear Equations, II

In this section we introduce the operation of pivoting and consider systems of linear equations which do not have unique solutions.

Roughly speaking, the Gaussian elimination method applied to a matrix proceeds as follows: Consider the columns one at a time, from left to right. For each column use the elementary row operations to transform the appropriate entry to a one and the remaining entries in the column to zeros. (The "appropriate" entry is the first entry in the first column, the second entry in the second column, and so forth.) This sequence of elementary row operations performed for each column is called *pivoting*. More precisely:

Method *To pivot a matrix about a given nonzero entry:*

1. Transform the given entry into a one.
2. Transform all other entries in the same column into zeros.

Pivoting is used in solving problems other than systems of linear equations. As we shall see in **Chapter 2** it is the basis for the simplex method of solving linear programming problems.

EXAMPLE 1 Pivot the matrix about the circled element.

$$\left[\begin{array}{cc|c} 18 & \textcircled{-6} & 15 \\ 5 & -2 & 4 \end{array}\right]$$

Solution The first step is to transform the -6 to a 1. We do this by multiplying the first row by $-\frac{1}{6}$:

$$\left[\begin{array}{cc|c} 18 & -6 & 15 \\ 5 & -2 & 4 \end{array}\right] \xrightarrow{-\frac{1}{6}[1]} \left[\begin{array}{cc|c} -3 & 1 & -\frac{5}{2} \\ 5 & -2 & 4 \end{array}\right].$$

Next, we transform the -2 (the only remaining entry in column 2) into a 0:

$$\left[\begin{array}{cc|c} -3 & 1 & -\frac{5}{2} \\ 5 & -2 & 4 \end{array}\right] \xrightarrow{[2]+2[1]} \left[\begin{array}{cc|c} -3 & 1 & -\frac{5}{2} \\ -1 & 0 & -1 \end{array}\right].$$

The last matrix is the result of pivoting the original matrix about the circled entry.

In terms of pivoting, we can give the following summary of the Gaussian elimination method.

Gaussian Elimination Method to Transform a System of Linear Equations into Diagonal Form

1. Write down the matrix corresponding to the linear system.
2. Make sure that the first entry in the first column is nonzero. Do this by interchanging the first row with one of the rows below it, if necessary.
3. Pivot the matrix about the first entry in the first column.
4. Make sure that the second entry in the second column is nonzero. Do this by interchanging the second row with one of the rows below it, if necessary.
5. Pivot the matrix about the second entry in the second column.
6. Continue in this manner.

All the systems considered in the preceding section had only a single solution. In this case we say that the solution is *unique*. Let us now use the Gaussian elimination method to study the various possibilities other than a unique solution. We first experiment with an example.

EXAMPLE 2 Determine all solutions of the system

$$\begin{cases} 2x + 2y + 4z = 8 \\ x - y + 2z = 2 \\ -x + 5y - 2z = 2. \end{cases}$$

Solution We set up the matrix corresponding to the system and perform the appropriate pivoting operations. (The elements pivoted about are circled.)

$$\left[\begin{array}{rrr|r} \textcircled{2} & 2 & 4 & 8 \\ 1 & -1 & 2 & 2 \\ -1 & 5 & -2 & 2 \end{array}\right] \xrightarrow[\substack{[2]+(-1)[1] \\ [3]+(1)[1]}]{\frac{1}{2}[1]} \left[\begin{array}{rrr|r} 1 & 1 & 2 & 4 \\ 0 & \textcircled{-2} & 0 & -2 \\ 0 & 6 & 0 & 6 \end{array}\right]$$

$$\xrightarrow[\substack{[1]+(-1)[2] \\ [3]+(-6)[2]}]{(-\frac{1}{2})[2]} \left[\begin{array}{rrr|r} 1 & 0 & 2 & 3 \\ 0 & 1 & 0 & 1 \\ 0 & 0 & 0 & 0 \end{array}\right].$$

Note that our method must terminate here, since there is no way to transform the third entry in the third column into a 1 without disturbing the columns already in appropriate form. The equations corresponding to the last matrix read

$$\begin{cases} x \quad + 2z = 3 \\ \quad y \quad = 1 \\ \quad 0 \quad = 0. \end{cases}$$

The last equation does not involve any of the variables and so may be omitted. This leaves the two equations

$$\begin{cases} x \quad + 2z = 3 \\ \quad y \quad = 1. \end{cases}$$

Now, taking the $2z$ term in the first equation to the right side, we can write the equations

$$\begin{cases} x = 3 - 2z \\ y = 1. \end{cases}$$

The value of y is given: $y = 1$. The value of x is given in terms of z. To find a solution to this system, assign any value to z. Then the first equation gives a value for x and thereby a specific solution to the system. For example, if we take $z = 1$, then the corresponding specific solution is

$$z = 1$$
$$x = 3 - 2(1) = 1$$
$$y = 1.$$

If we take $z = -3$, the corresponding specific solution is

$$z = -3$$
$$x = 3 - 2(-3) = 9$$
$$y = 1.$$

Thus, we see that the original system has infinitely many specific solutions, corresponding to the infinitely many possible different choices for z.

We say that the *general solution* of the system is

$$\begin{aligned} z &= \text{any value} \\ x &= 3 - 2z \\ y &= 1. \end{aligned}$$

When a linear system cannot be *completely* diagonalized:

1. Apply the Gaussian elimination method to as many columns as possible. Proceed from left to right, but do not disturb columns that have already been put into proper form.
2. Variables corresponding to columns not in proper form can assume any value.
3. The other variables can be expressed in terms of the variables of step 2.

EXAMPLE 3 Find all solutions of the linear system

$$\begin{cases} x + 2y - z + 3w = 5 \\ \quad\;\; y + 2z + w = 7. \end{cases}$$

Solution The Gaussian elimination method proceeds as follows:

$$\left[\begin{array}{cccc|c} 1 & 2 & -1 & 3 & 5 \\ 0 & \textcircled{1} & 2 & 1 & 7 \end{array}\right] \quad \text{(The first column is already in proper form.)}$$

$$\xrightarrow{[1] + (-2)[2]} \left[\begin{array}{cccc|c} 1 & 0 & -5 & 1 & -9 \\ 0 & 1 & 2 & 1 & 7 \end{array}\right].$$

We cannot do anything further with the last two columns (without disturbing the first two columns), so the corresponding variables, z and w, can assume any values. Writing down the equations corresponding to the last matrix yields

$$\begin{cases} x \qquad - 5z + w = -9 \\ \quad\;\; y + 2z + w = 7 \end{cases}$$

or

$$\begin{aligned} z &= \text{any value} \\ w &= \text{any value} \\ x &= -9 + 5z - w \\ y &= 7 - 2z - w. \end{aligned}$$

To determine an example of a specific solution, let $z = 1$, $w = 2$. Then a specific solution of the original system is

$$
\begin{aligned}
z &= 1 \\
w &= 2 \\
x &= -9 + 5(1) - (2) = -6 \\
y &= \ \ 7 - 2(1) - (2) = \ \ 3.
\end{aligned}
$$

EXAMPLE 4 Find all solutions of the system of equations

$$
\begin{cases} x - 7y + z = 3 \\ 2x - 14y + 3z = 4. \end{cases}
$$

Solution The first pivot operation is routine:

$$
\left[\begin{array}{ccc|c} \textcircled{1} & -7 & 1 & 3 \\ 2 & -14 & 3 & 4 \end{array}\right] \xrightarrow{[2] + (-2)[1]} \left[\begin{array}{ccc|c} 1 & -7 & 1 & 3 \\ 0 & 0 & 1 & -2 \end{array}\right].
$$

However, it is impossible to pivot about the zero in the second column. So skip the second column and pivot about the second entry in the third column to get

$$
\left[\begin{array}{ccc|c} 1 & -7 & 0 & 5 \\ 0 & 0 & 1 & -2 \end{array}\right].
$$

This is as far as we can go. The variable corresponding to the second column, namely y, can assume any value, and the general solution of the system is obtained from the equations

$$
\begin{cases} x - 7y \qquad = 5 \\ \qquad\qquad z = -2. \end{cases}
$$

Therefore, the general solution of the system is

$$
\begin{aligned}
y &= \text{any value} \\
x &= 5 + 7y \\
z &= -2.
\end{aligned}
$$

We have seen that a linear system may have a unique solution or it may have infinitely many solutions. But another phenomenon can occur: A system may have no solutions at all, as the next example shows.

EXAMPLE 5 Find all solutions of the system

$$
\begin{cases} x - y + z = 3 \\ x + y - z = 5 \\ -2x + 4y - 4z = 1. \end{cases}
$$

Solution We apply the Gaussian elimination method to the matrix of the system.

$$\left[\begin{array}{rrr|r} \textcircled{1} & -1 & 1 & 3 \\ 1 & 1 & -1 & 5 \\ -2 & 4 & -4 & 1 \end{array}\right] \xrightarrow[{[3] + 2[1]}]{[2] + (-1)[1]} \left[\begin{array}{rrr|r} 1 & -1 & 1 & 3 \\ 0 & \textcircled{2} & -2 & 2 \\ 0 & 2 & -2 & 7 \end{array}\right]$$

$$\xrightarrow[{[3] + (-2)[2]}]{\substack{\frac{1}{2}[2] \\ [1] + (1)[2]}} \left[\begin{array}{rrr|r} 1 & 0 & 0 & 4 \\ 0 & 1 & -1 & 1 \\ 0 & 0 & 0 & 5 \end{array}\right].$$

We cannot pivot about the last zero in the third column, so we have carried the method as far as we can. Let us write out the equations corresponding to the last matrix:

$$\begin{cases} x \qquad\quad = 4 \\ \quad y - z = 1 \\ \quad 0 \qquad = 5. \end{cases}$$

Note that the last equation is a built-in contradiction. In mathematical terms, the last equation is *inconsistent*. So the last equation can never be satisfied, no matter what the values of x, y, z. Thus, the original system has no solutions. Systems with no solutions can always be detected by the presence of inconsistent equations in the last matrix resulting from the Gaussian elimination method.

At first it might seem strange that some systems have no solutions, some have one, and yet others have infinitely many. The reason for the difference can be explained geometrically. For simplicity, consider the case of systems of two equations in two variables. Each equation in this case has a graph in the xy-plane, and the graph is a straight line. As we have seen, solving the system corresponds to finding the points lying on both lines. There are three possibilities. First, the two lines may intersect. In this case the solution is unique. Second, the two lines may be parallel. Then the two lines do not intersect and the system has no solutions. Finally, the two equations may represent the same line, as, for example, do the equations $2x + 3y = 1$, $4x + 6y = 2$. In this case every point on the line is a solution of the system; that is, there are infinitely many solutions (Fig. 1).

FIGURE 1

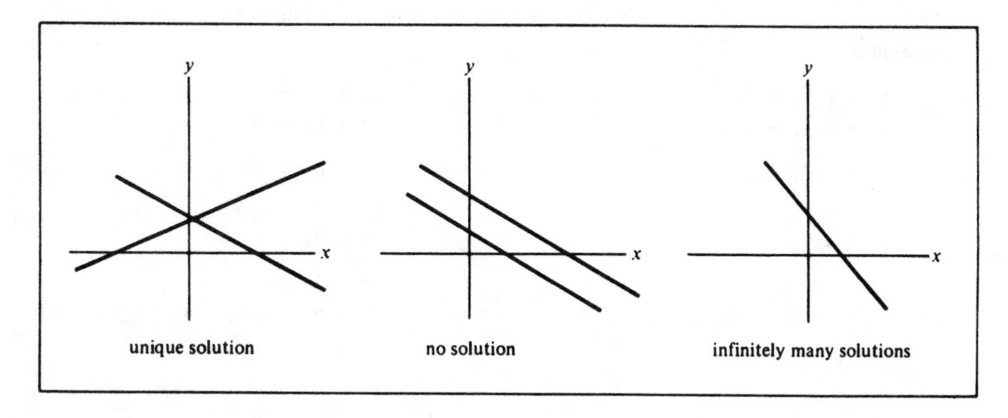

PRACTICE PROBLEMS 2

1. Find a specific solution to a system of linear equations whose general solution is

$$\begin{aligned} w &= \text{any value} \\ y &= \text{any value} \\ z &= 7 + 6w \\ x &= 26 - 2y + 14w. \end{aligned}$$

2. Find all solutions of this system of linear equations.

$$\begin{cases} 2x + 4y - 4z - 4w = 24 \\ -3x - 6y + 10z - 18w = -8 \\ -x - 2y + 4z - 10w = 2. \end{cases}$$

EXERCISES 2

Pivot each of the following matrices about the circled element.

1. $\begin{bmatrix} \textcircled{2} & -4 & 6 \\ 3 & 7 & 1 \end{bmatrix}$

2. $\begin{bmatrix} 1 & 2 & 3 \\ 4 & \textcircled{8} & -12 \end{bmatrix}$

3. $\begin{bmatrix} 7 & 1 & 4 & 5 \\ -1 & 1 & \textcircled{2} & 6 \\ 4 & 0 & 2 & 3 \end{bmatrix}$

4. $\begin{bmatrix} 5 & 10 & -10 & 12 \\ 4 & 3 & 6 & 12 \\ 4 & \textcircled{-4} & 4 & -16 \end{bmatrix}$

5. $\begin{bmatrix} \textcircled{2} & 3 \\ 6 & 0 \\ 1 & 5 \end{bmatrix}$

6. $\begin{bmatrix} 2 & 1 \\ \textcircled{-1} & 0 \end{bmatrix}$

7. $\begin{bmatrix} 4 & 3 & 0 \\ \frac{2}{3} & 0 & -2 \\ 1 & 3 & \textcircled{6} \end{bmatrix}$

8. $\begin{bmatrix} 1 & 0 & 2 \\ -1 & 1 & \textcircled{-2} \\ 1 & 2 & 6 \end{bmatrix}$

Use the Gaussian elimination method to find all solutions of the following systems of linear equations.

9. $\begin{cases} 2x - 4y = 6 \\ -x + 2y = -3 \end{cases}$

10. $\begin{cases} -\frac{1}{2}x + y = \frac{3}{2} \\ -3x + 6y = 10 \end{cases}$

11. $\begin{cases} x + 2y = 5 \\ 3x - y = 1 \\ -x + 3y = 5 \end{cases}$

12. $\begin{cases} x - 6y = 12 \\ -\frac{1}{2}x + 3y = -6 \\ \frac{1}{3}x - 2y = 4 \end{cases}$

13. $\begin{cases} x - y + 3z = 3 \\ -2x + 3y - 11z = -4 \\ x - 2y + 8z = 6 \end{cases}$

14. $\begin{cases} x - 3y + z = 5 \\ -2x + 7y - 6z = -9 \\ x - 2y - 3z = 6 \end{cases}$

15. $\begin{cases} x + y + z = -1 \\ 2x + 3y + 2z = 3 \\ 2x + y + 2z = -7 \end{cases}$

16. $\begin{cases} x - 3y + 2z = 10 \\ -x + 3y - z = -6 \\ -x + 3y + 2z = 6 \end{cases}$

17. $\begin{cases} x + 2y + 3z = 4 \\ 5x + 6y + 7z = 8 \\ x + 2y + 3z = 5 \end{cases}$

18. $\begin{cases} x + 3y = 7 \\ x + 2y = 5 \\ -x + y = 2 \end{cases}$

19. $\begin{cases} x + y - 2z + 2w = 5 \\ 2x + y - 4z + w = 5 \\ 3x + 4y - 6z + 9w = 20 \\ 4x + 4y - 8z + 8w = 20 \end{cases}$

20. $\begin{cases} 2y + z - w = 1 \\ x - y + z + w = 14 \\ -x - 9y - z + 4w = 11 \\ x + y + z = 9 \end{cases}$

21. In a laboratory experiment, a researcher wants to provide a rabbit with exactly 1000 units of vitamin A, exactly 1600 units of vitamin C, and exactly 2400 units of vitamin E. The rabbit is fed a mixture of three foods. Each gram of food 1 contains 2 units of vitamin A, 3 units of vitamin C, and 5 units of vitamin E. Each gram of food 2 contains 4 units of vitamin A, 7 units of vitamin C, and 9 units of vitamin E. Each gram of food 3 contains 6 units of vitamin A, 10 units of vitamin C, and 14 units of vitamin E. How many grams of each food should the rabbit be fed?

22. Rework Exercise 21 with the requirement for vitamin E changed to 2000 units.

SOLUTIONS TO PRACTICE PROBLEMS 2

1. Since w and y can each assume any value, select any numbers, say $w = 1$ and $y = 2$. Then $z = 7 + 6(1) = 13$ and $x = 26 - 2(2) + 14(1) = 36$. So $x = 36$, $y = 2$, $z = 13$, $w = 1$ is a specific solution. There are infinitely many different specific solutions, since there are infinitely many different choices for w and y.

2. We apply the Gaussian elimination method to the matrix of the system.

$$\left[\begin{array}{cccc|c} \textcircled{2} & 4 & -4 & -4 & 24 \\ -3 & -6 & 10 & -18 & -8 \\ -1 & -2 & 4 & -10 & 2 \end{array}\right] \xrightarrow[{[3] + 1[1]}]{\frac{1}{2}[1] \quad [2] + 3[1]} \left[\begin{array}{cccc|c} 1 & 2 & -2 & -2 & 12 \\ 0 & 0 & \textcircled{4} & -24 & 28 \\ 0 & 0 & 2 & -12 & 14 \end{array}\right]$$

$$\xrightarrow[{[3] + (-2)[2]}]{\frac{1}{4}[2] \quad [1] + 2[2]} \left[\begin{array}{cccc|c} 1 & 2 & 0 & -14 & 26 \\ 0 & 0 & 1 & -6 & 7 \\ 0 & 0 & 0 & 0 & 0 \end{array}\right].$$

The corresponding system of equation is

$$\begin{cases} x + 2y \qquad - 14w = 26 \\ \qquad\qquad z - 6w = 7. \end{cases}$$

The general solution is

$$\begin{aligned} w &= \text{any value} \\ y &= \text{any value} \\ z &= 7 + 6w \\ x &= 26 - 2y + 14w. \end{aligned}$$

1.3 Arithmetic Operations on Matrices

We introduced matrices in the preceding sections in order to display the coefficients of a system of linear equations. For example, the linear system

$$\begin{cases} 5x - 3y = \frac{1}{2} \\ 4x + 2y = -1 \end{cases}$$

is represented by the matrix

$$\left[\begin{array}{cc|c} 5 & -3 & \frac{1}{2} \\ 4 & 2 & -1 \end{array}\right].$$

After we have become accustomed to using such matrices in solving linear systems, we may omit the vertical line which separates the left and the right sides of the equations. We need only remember that the right side of the equations is recorded in the right column. So, for example, we would write the matrix above in the form

$$\begin{bmatrix} 5 & -3 & \frac{1}{2} \\ 4 & 2 & -1 \end{bmatrix}.$$

A matrix is *any* rectangular array of numbers and may be of any size. Here are some examples of matrices of various sizes:

$$\begin{bmatrix} 3 & 7 \\ 0 & -1 \end{bmatrix}, \quad \begin{bmatrix} 1 \\ 2 \end{bmatrix}, \quad \begin{bmatrix} 2 & 1 \end{bmatrix}, \quad [6], \quad \begin{bmatrix} 5 & 7 & -1 \\ 0 & 3 & 5 \\ 6 & 0 & 5 \end{bmatrix}.$$

Examples of matrices abound in everyday life. For example, the newspaper stock-market report is a large matrix with several thousand rows, one for each listed stock. The columns of the matrix give various data about each stock, such as opening and closing price, number of shares traded, and so forth. Another example of a matrix is a mileage charge on a road map. The rows and columns are labeled with the names of cities. The number at a given row and column gives the distance between the corresponding cities.

In these everyday examples, matrices are used only to display data. However, the most important applications involve arithmetic operations on matrices—namely, addition, subtraction, and multiplication of matrices. The major goal of this section is to discuss these operations. Before we can do so, however, we need some vocabulary with which to describe matrices.

A matrix is described by the number of rows and columns it contains. For example, the matrix

$$\begin{bmatrix} 7 & 5 \\ \frac{1}{2} & -2 \\ 2 & -11 \end{bmatrix}$$

has three rows and two columns and is referred to as a 3×2 (read: "three-by-two") *matrix*. The matrix $[4 \quad 5 \quad 0]$ has one row and three columns and is a 1×3

matrix. A matrix with only one row is often called a *row matrix* (sometimes also called a *row vector*). A matrix, such as $\begin{bmatrix} 2 \\ 7 \end{bmatrix}$, which has only one column is called a *column matrix* or *column vector*. If a matrix has the same number of rows and columns, it is called a *square matrix*. Here are some square matrices of various sizes:

$$[5], \quad \begin{bmatrix} 1 & 2 \\ 3 & 4 \end{bmatrix}, \quad \begin{bmatrix} 2 & -1 & 0 \\ 3 & 5 & 4 \\ 0 & 3 & -7 \end{bmatrix}.$$

The rows of a matrix are numbered from the top down, and the columns are numbered from left to right. For example, the first row of the matrix

$$\begin{bmatrix} 1 & -1 & 0 \\ 2 & 1 & 7 \\ -3 & 2 & 4 \end{bmatrix}$$

is $[1 \quad -1 \quad 0]$, and its third column is

$$\begin{bmatrix} 0 \\ 7 \\ 4 \end{bmatrix}.$$

The numbers in a matrix, called *entries*, may be identified in terms of the row and column containing the entry in question. For example, the entry in the first row, third column of the following matrix is 0:

$$\begin{bmatrix} 1 & -1 & \mathbf{0} \\ 2 & 1 & 7 \\ -3 & 2 & 4 \end{bmatrix};$$

the entry in the second row, first column is 2:

$$\begin{bmatrix} 1 & -1 & 0 \\ \mathbf{2} & 1 & 7 \\ -3 & 2 & 4 \end{bmatrix};$$

and the entry in the third row, third column is 4:

$$\begin{bmatrix} 1 & -1 & 0 \\ 2 & 1 & 7 \\ -3 & 2 & \mathbf{4} \end{bmatrix}.$$

We use the double-subscripted letters to indicate the locations of the entries of a matrix. We denote the entry in the ith row, jth column by a_{ij}. For instance, in the above matrix we have $a_{13} = 0$, $a_{21} = 2$, and $a_{33} = 4$.

We say that two matrices A and B are *equal*, denoted $A = B$, provided that they have the same size and that all their corresponding entries are equal.

Addition and Subtraction of Matrices We define the sum $A + B$ of two matrices A and B only if A and B are two matrices of the same size—that is, if A and B

have the same number of rows and the same number of columns. In this case $A + B$ is the matrix formed by adding the corresponding entries of A and B. Thus, for example,

$$\begin{bmatrix} 2 & 0 \\ 1 & 1 \\ 5 & 3 \end{bmatrix} + \begin{bmatrix} 5 & 4 \\ 0 & 2 \\ 2 & 6 \end{bmatrix} = \begin{bmatrix} 2+5 & 0+4 \\ 1+0 & 1+2 \\ 5+2 & 3+6 \end{bmatrix} = \begin{bmatrix} 7 & 4 \\ 1 & 3 \\ 7 & 9 \end{bmatrix}.$$

We subtract matrices of the same size by subtracting corresponding entries. Thus, we have

$$\begin{bmatrix} 7 \\ 1 \end{bmatrix} - \begin{bmatrix} 3 \\ 2 \end{bmatrix} = \begin{bmatrix} 7-3 \\ 1-2 \end{bmatrix} = \begin{bmatrix} 4 \\ -1 \end{bmatrix}.$$

Multiplication of Matrices It might seem that to define the product of two matrices, one would start with two matrices of like size and multiply the corresponding entries. But this definition is not useful, since the calculations that arise in applications require a somewhat more complex multiplication. In the interests of simplicity, we start by defining the product of a row matrix times a column matrix.

If A is a row matrix and B is a column matrix, then we can form the product $A \cdot B$ provided that the two matrices have the same length. The product $A \cdot B$ is the 1×1 matrix obtained by multiplying corresponding entries of A and B and then forming the sum.

We may put this definition into algebraic terms as follows. Suppose that A is the row matrix

$$A = [a_1 \quad a_2 \quad \cdots \quad a_n],$$

and B is the column matrix

$$B = \begin{bmatrix} b_1 \\ b_2 \\ \vdots \\ b_n \end{bmatrix}.$$

Note that A and B are both of the same length, namely n. Then

$$A \cdot B = [a_1 \quad a_2 \quad \cdots \quad a_n] \cdot \begin{bmatrix} b_1 \\ b_2 \\ \vdots \\ b_n \end{bmatrix}$$

is calculated by multiplying corresponding entries of A and B and forming the sum; that is,

$$A \cdot B = [a_1b_1 + a_2b_2 + \cdots + a_nb_n].$$

Notice that the product is a 1×1 matrix, namely a single number in brackets.

Here are some examples of the product of a row matrix times a column matrix:

$$[3 \quad \tfrac{1}{2}] \cdot \begin{bmatrix} 1 \\ 4 \end{bmatrix} = [3 \cdot 1 + \tfrac{1}{2} \cdot 4] = [5],$$

$$[2 \quad 0 \quad -1] \cdot \begin{bmatrix} 6 \\ 5 \\ 3 \end{bmatrix} = [2 \cdot 6 + 0 \cdot 5 + (-1) \cdot 3] = [9].$$

In multiplying a row matrix times a column matrix, it helps to use both hands. Use your left index finger to point to an element of the row matrix and your right to point to the corresponding element of the column. Multiply the elements you are pointing to and keep a running total of the products in your head. After each multiplication move your fingers to the next elements of each matrix. With a little practice you should be able to multiply a row times a column quickly and accurately.

The definition of multiplication above may seem strange. But products of this sort occur in many down-to-earth problems. Consider, for instance, the next example.

EXAMPLE 1 A dairy farm produces three items—milk, eggs, and cheese. The prices of these items are $1.50 per gallon, $.80 per dozen, and $2 per pound, respectively. In a certain week the dairy farm sells 30,000 gallons of milk, 2000 dozen eggs, and 5000 pounds of cheese. Represent its total revenue as a matrix product.

Solution The total revenue equals

$$(1.50)(30{,}000) + (.80)(2000) + (2)(5000).$$

This suggests that we define two matrices: the first displays the prices of the various produce:

$$[1.50 \quad .80 \quad 2].$$

The second represents the production:

$$\begin{bmatrix} 30{,}000 \\ 2000 \\ 5000 \end{bmatrix}.$$

Then the revenue for the week, when placed in a 1×1 matrix, equals

$$[1.50 \quad .80 \quad 2] \begin{bmatrix} 30{,}000 \\ 2000 \\ 5000 \end{bmatrix} = [56{,}600].$$

The principle behind Example 1 is this: any sum of products of the form $a_1b_1 + a_2b_2 + \cdots + a_nb_n$, when placed in a 1×1 matrix, can be written as the matrix product

$$[a_1b_1 + a_2b_2 + \cdots + a_nb_n] = [a_1 \quad a_2 \quad \cdots \quad a_n]\begin{bmatrix} b_1 \\ b_2 \\ \vdots \\ b_n \end{bmatrix}.$$

Let us illustrate the procedure for multiplying more general matrices by working out a typical product:

$$\begin{bmatrix} 2 & 1 \\ 0 & 1 \\ 1 & 0 \end{bmatrix} \cdot \begin{bmatrix} 1 & 1 \\ 4 & 2 \end{bmatrix}.$$

To obtain the entries of the product, we multiply the rows of the left matrix by the columns of the right matrix, taking care to arrange the products in a specific way to yield a matrix, as follows. Start with the first row on the left, $[2 \quad 1]$, and the first column on the right, $\begin{bmatrix} 1 \\ 4 \end{bmatrix}$. Their product is $[6]$, so we enter 6 as the element in the first row, first column of the product:

$$\begin{bmatrix} \mathbf{2} & \mathbf{1} \\ 0 & 1 \\ 1 & 0 \end{bmatrix} \cdot \begin{bmatrix} \mathbf{1} & 1 \\ \mathbf{4} & 2 \end{bmatrix} = \begin{bmatrix} \mathbf{6} & \\ & \\ & \end{bmatrix}.$$

The product of the first row of the left matrix and the second column of the right matrix is $[4]$, so we put a 4 in the first row, second column of the product:

$$\begin{bmatrix} 2 & 1 \\ 0 & 1 \\ 1 & 0 \end{bmatrix} \cdot \begin{bmatrix} 1 & 1 \\ 4 & 2 \end{bmatrix} = \begin{bmatrix} 6 & 4 \\ & \\ & \end{bmatrix}.$$

There are no more columns which can be multiplied by the first row, so let us move to the second row and shift back to the first column. Correspondingly, we move one row down in the product:

$$\begin{bmatrix} 2 & 1 \\ 0 & 1 \\ 1 & 0 \end{bmatrix} \cdot \begin{bmatrix} 1 & 1 \\ 4 & 2 \end{bmatrix} = \begin{bmatrix} 6 & 4 \\ 4 & \\ & \end{bmatrix}$$

$$\begin{bmatrix} 2 & 1 \\ 0 & 1 \\ 1 & 0 \end{bmatrix} \cdot \begin{bmatrix} 1 & 1 \\ 4 & 2 \end{bmatrix} = \begin{bmatrix} 6 & 4 \\ 4 & 2 \\ & \end{bmatrix}.$$

We have now exhausted the second row of the left matrix, so we shift to the third row and correspondingly move down one row in the product.

$$\begin{bmatrix} 2 & 1 \\ 0 & 1 \\ 1 & 0 \end{bmatrix} \cdot \begin{bmatrix} 1 & 1 \\ 4 & 2 \end{bmatrix} = \begin{bmatrix} 6 & 4 \\ 4 & 2 \\ 1 & \end{bmatrix}$$

$$\begin{bmatrix} 2 & 1 \\ 0 & 1 \\ 1 & 0 \end{bmatrix} \cdot \begin{bmatrix} 1 & 1 \\ 4 & 2 \end{bmatrix} = \begin{bmatrix} 6 & 4 \\ 4 & 2 \\ 1 & 1 \end{bmatrix}.$$

Note that we have now multiplied every row of the left matrix by every column of the right matrix. This completes computation of the product:

$$\begin{bmatrix} 2 & 1 \\ 0 & 1 \\ 1 & 0 \end{bmatrix} \cdot \begin{bmatrix} 1 & 1 \\ 4 & 2 \end{bmatrix} = \begin{bmatrix} 6 & 4 \\ 4 & 2 \\ 1 & 1 \end{bmatrix}.$$

EXAMPLE 2 Calculate the following product:

$$\begin{bmatrix} 1 & 5 \\ 3 & 2 \end{bmatrix} \cdot \begin{bmatrix} 1 & 2 \\ 1 & 0 \end{bmatrix}.$$

Solution

$$\begin{bmatrix} 1 & 5 \\ 3 & 2 \end{bmatrix} \cdot \begin{bmatrix} 1 & 2 \\ 1 & 0 \end{bmatrix} = \begin{bmatrix} 6 & \\ & \end{bmatrix}$$

$$\begin{bmatrix} 1 & 5 \\ 3 & 2 \end{bmatrix} \cdot \begin{bmatrix} 1 & 2 \\ 1 & 0 \end{bmatrix} = \begin{bmatrix} 6 & 2 \\ & \end{bmatrix}$$

$$\begin{bmatrix} 1 & 5 \\ 3 & 2 \end{bmatrix} \cdot \begin{bmatrix} 1 & 2 \\ 1 & 0 \end{bmatrix} = \begin{bmatrix} 6 & 2 \\ 5 & \end{bmatrix}$$

$$\begin{bmatrix} 1 & 5 \\ 3 & 2 \end{bmatrix} \cdot \begin{bmatrix} 1 & 2 \\ 1 & 0 \end{bmatrix} = \begin{bmatrix} 6 & 2 \\ 5 & 6 \end{bmatrix}$$

Thus

$$\begin{bmatrix} 1 & 5 \\ 3 & 2 \end{bmatrix} \cdot \begin{bmatrix} 1 & 2 \\ 1 & 0 \end{bmatrix} = \begin{bmatrix} 6 & 2 \\ 5 & 6 \end{bmatrix}.$$

Notice that we cannot use the method above to compute the product $A \cdot B$ of *any* matrices A and B. For the procedure to work, it is crucial that the number of entries of each row of A be the same as the number of entries of each column of B. (Or, to put it another way, the number of columns of the left matrix must equal the number of rows of the right matrix.) Therefore, in order for us to form the product $A \cdot B$, the sizes of A and B must match up in a special way. If A is $m \times n$ and B is $p \times q$, then the product $A \cdot B$ is defined only in case the "inner" dimensions n and p are equal. In that case, the product is determined by the "outer" dimensions m and q. It is an $m \times q$ matrix.

$$\underset{m \times n}{A} \cdot \underset{p \times q}{B} = \underset{m \times q}{C}.$$

(n and p: equal)

So, for example,

$$\underset{3 \times 4}{\begin{bmatrix} & & & \\ & & & \\ & & & \end{bmatrix}} \underset{4 \times 2}{\begin{bmatrix} & \\ & \\ & \\ & \end{bmatrix}} = \underset{3 \times 2}{\begin{bmatrix} & \\ & \\ & \end{bmatrix}}$$

$$\underset{2 \times 2}{\begin{bmatrix} & \\ & \end{bmatrix}} \underset{2 \times 1}{\begin{bmatrix} \\ \end{bmatrix}} = \underset{2 \times 1}{\begin{bmatrix} \\ \end{bmatrix}}.$$

If the sizes of A and B do not match up in the way just described, the product $A \cdot B$ is not defined.

EXAMPLE 3 Calculate the following products, if defined.

(a) $\begin{bmatrix} 3 & -1 \\ 2 & 0 \\ 1 & 5 \end{bmatrix} \begin{bmatrix} 1 & 0 \\ 5 & -4 \\ 2 & -1 \end{bmatrix}$ (b) $\begin{bmatrix} 3 & -1 \\ 2 & 0 \\ 1 & 5 \end{bmatrix} \begin{bmatrix} 5 & 4 \\ -2 & 3 \end{bmatrix}$

Solution (a) The matrices to be multiplied are 3×2 and 3×2. The inner dimensions do not match, so the product is undefined.

(b) We are asked to multiply a 3×2 matrix times a 2×2 matrix. The inner dimensions match, so the product is defined and has size determined by the outer dimensions, that is, 3×2.

$$\begin{bmatrix} 3 & -1 \\ 2 & 0 \\ 1 & 5 \end{bmatrix} \cdot \begin{bmatrix} 5 & 4 \\ -2 & 3 \end{bmatrix} = \begin{bmatrix} 3 \cdot 5 + (-1) \cdot (-2) & 3 \cdot 4 + (-1) \cdot 3 \\ 2 \cdot 5 + 0 \cdot (-2) & 2 \cdot 4 + 0 \cdot 3 \\ 1 \cdot 5 + 5 \cdot (-2) & 1 \cdot 4 + 5 \cdot 3 \end{bmatrix}$$
$$= \begin{bmatrix} 17 & 9 \\ 10 & 9 \\ -5 & 19 \end{bmatrix}.$$

Multiplication of matrices has many properties in common with multiplication of ordinary numbers. However, there is at least one important difference. With matrix multiplication, the order of the factors is usually important. For example, the product of a 2×3 matrix times a 3×2 matrix is defined and the product is a 2×2 matrix. If the order is reversed to a 3×2 matrix times a 2×3 matrix, the product is a 3×3 matrix. So reversing the order may change the size of the product. Even when it does not, reversing the order may still change the entries in the product, as the following two products demonstrate.

$$\begin{bmatrix} 1 & 5 \\ 3 & 2 \end{bmatrix} \begin{bmatrix} 1 & 2 \\ 1 & 0 \end{bmatrix} = \begin{bmatrix} 6 & 2 \\ 5 & 6 \end{bmatrix}$$

$$\begin{bmatrix} 1 & 2 \\ 1 & 0 \end{bmatrix} \begin{bmatrix} 1 & 5 \\ 3 & 2 \end{bmatrix} = \begin{bmatrix} 7 & 9 \\ 1 & 5 \end{bmatrix}.$$

EXAMPLE 4 An investment trust has investments in three states. Its deposits in each state are divided among bonds, mortgages, and consumer loans. On January 1 the amount (in millions of dollars) of money invested in each category by state is given by the matrix

$$\begin{array}{l} \\ \\ \text{State A} \\ \text{State B} \\ \text{State C} \end{array} \begin{array}{c} \begin{array}{ccc} & & \text{Consumer} \\ \text{Bonds} & \text{Mortgages} & \text{loans} \end{array} \\ \begin{bmatrix} 10 & 5 & 20 \\ 30 & 12 & 10 \\ 15 & 6 & 25 \end{bmatrix} \end{array}.$$

The current average yields are 7% for bonds, 9% for mortgages, and 15% for consumer loans. Determine the earnings of the trust from its investments in each state.

Solution Define the matrix of investment yields by

$$\begin{bmatrix} .07 \\ .09 \\ .15 \end{bmatrix} \begin{array}{l} \text{Bonds} \\ \text{Mortgages} \\ \text{Consumer loans.} \end{array}$$

The amount earned in state A, for instance, is

$$\begin{aligned} &[\text{amount of bonds}][\text{yield of bonds}] \\ &\quad + [\text{amount of mortgages}][\text{yield of mortgages}] \\ &\quad + [\text{amount of consumer loans}][\text{yield of consumer loans}] \\ &= (10)(.07) + (5)(.09) + (20)(.15). \end{aligned}$$

And this is just the first entry of the product

$$\begin{bmatrix} \mathbf{10} & \mathbf{5} & \mathbf{20} \\ 30 & 12 & 10 \\ 15 & 6 & 25 \end{bmatrix} \begin{bmatrix} \mathbf{.07} \\ \mathbf{.09} \\ \mathbf{.15} \end{bmatrix}.$$

Similarly, the earnings for the other states are the second and third entries of the product. Carrying out the arithmetic, we find that

$$\begin{bmatrix} 10 & 5 & 20 \\ 30 & 12 & 10 \\ 15 & 6 & 25 \end{bmatrix} \begin{bmatrix} .07 \\ .09 \\ .15 \end{bmatrix} = \begin{bmatrix} 4.15 \\ 4.68 \\ 5.34 \end{bmatrix}.$$

Therefore, the trust earns $4.15 million in state A, $4.68 million in state B, and $5.34 million in state C.

EXAMPLE 5 A clothing manufacturer has factories in Los Angeles, San Antonio, and Newark. Sales (in thousands) during the first quarter of last year are summarized in the production matrix

$$\begin{array}{l} \\ \text{Coats} \\ \text{Shirts} \\ \text{Sweaters} \\ \text{Ties} \end{array} \begin{array}{c} \begin{array}{ccc} \text{Los Angeles} & \text{San Antonio} & \text{Newark} \end{array} \\ \begin{bmatrix} 12 & 13 & 38 \\ 25 & 5 & 26 \\ 11 & 8 & 8 \\ 5 & 0 & 12 \end{bmatrix} \end{array}.$$

During this period the selling price of a coat was $100, of a shirt $10, of a sweater $25, and of a tie $5.

(a) Use a matrix calculation to determine the total revenue produced by each of the factories.

(b) Suppose that the prices had been $110, $8, $20, and $10, respectively. How would this have affected the revenue of each factory?

Solution (a) For each factory, we wish to multiply the price of each item by the number produced to arrive at revenue. Since the production figures for the various items of clothing are arranged down the columns, we arrange the prices in a row matrix, ready for multiplication; the price matrix is

$$[100 \quad 10 \quad 25 \quad 5].$$

The revenues of the various factories are then the entries of the product

$$[100 \quad 10 \quad 25 \quad 5]\begin{bmatrix} 12 & 13 & 38 \\ 25 & 5 & 26 \\ 11 & 8 & 8 \\ 5 & 0 & 12 \end{bmatrix} = \begin{array}{ccc} \text{Los Angeles} & \text{San Antonio} & \text{Newark} \\ [1750 & 1550 & 4320]. \end{array}$$

Since the production figures are in thousands, the revenue figures are in thousands of dollars. That is, the Los Angeles factory has revenues of $1,750,000, the San Antonio factory $1,550,000, and the Newark factory $4,320,000.

(b) In a similar way, we determine the revenue of each factory if the price matrix had been $[110 \quad 8 \quad 20 \quad 10]$.

$$[110 \quad 8 \quad 20 \quad 10]\begin{bmatrix} 12 & 13 & 38 \\ 25 & 5 & 26 \\ 11 & 8 & 8 \\ 5 & 0 & 12 \end{bmatrix} = \begin{array}{ccc} \text{Los Angeles} & \text{San Antonio} & \text{Newark} \\ [1790 & 1630 & 4668]. \end{array}$$

The change in revenue at each factory can be read from the difference of the revenue matrices:

$$[1790 \quad 1630 \quad 4668] - [1750 \quad 1550 \quad 4320] = [40 \quad 80 \quad 348].$$

That is, if prices had been as given in part (b), then revenues of the Los Angeles factory would have increased by 40, revenues at San Antonio would have increased by 80, and revenues at Newark would have increased by 348.

There are special matrices analogous to the number 1. Such matrices are called *identity matrices.* The identity matrix I_n of size n is the $n \times n$ square matrix with all zeros, except for ones down the upper-left-to-lower-right diagonal. Here are the identity matrices of sizes 2, 3, and 4:

$$I_2 = \begin{bmatrix} 1 & 0 \\ 0 & 1 \end{bmatrix}, \quad I_3 = \begin{bmatrix} 1 & 0 & 0 \\ 0 & 1 & 0 \\ 0 & 0 & 1 \end{bmatrix}, \quad I_4 = \begin{bmatrix} 1 & 0 & 0 & 0 \\ 0 & 1 & 0 & 0 \\ 0 & 0 & 1 & 0 \\ 0 & 0 & 0 & 1 \end{bmatrix}.$$

The characteristic property of an identity matrix is that it plays the role of the number 1; that is,

$$I_n \cdot A = A \cdot I_n = A$$

for all $n \times n$ matrices A.

One of the principal uses of matrices is in dealing with systems of linear equations. Matrices provide a compact way of writing systems, as the next example shows.

EXAMPLE 6 Write the system of linear equations

$$\begin{cases} -2x + 4y = 2 \\ -3x + 7y = 7 \end{cases}$$

as a matrix equation.

Solution The system of equations can be written in the form

$$\begin{bmatrix} -2x + 4y \\ -3x + 7y \end{bmatrix} = \begin{bmatrix} 2 \\ 7 \end{bmatrix}.$$

So consider the matrices

$$A = \begin{bmatrix} -2 & 4 \\ -3 & 7 \end{bmatrix}, \quad X = \begin{bmatrix} x \\ y \end{bmatrix}, \quad B = \begin{bmatrix} 2 \\ 7 \end{bmatrix}.$$

Notice that

$$AX = \begin{bmatrix} -2 & 4 \\ -3 & 7 \end{bmatrix} \begin{bmatrix} x \\ y \end{bmatrix} = \begin{bmatrix} -2x + 4y \\ -3x + 7y \end{bmatrix}.$$

Thus AX is a 2×1 column matrix whose entries correspond to the left side of the given system of linear equations. Since the entries of B correspond to the right side of the system of equations, we can rewrite the given system in the form

$$AX = B$$

—that is,

$$\begin{bmatrix} -2 & 4 \\ -3 & 7 \end{bmatrix} \begin{bmatrix} x \\ y \end{bmatrix} = \begin{bmatrix} 2 \\ 7 \end{bmatrix}.$$

The matrix A of the example above displays the coefficients of the variables x and y, so it is called the *coefficient matrix* of the system.

PRACTICE PROBLEMS 3

1. Compute

$$\begin{bmatrix} 3 & 1 & 2 \\ -1 & 0 & \frac{1}{2} \\ 0 & 4 & 1 \end{bmatrix}\begin{bmatrix} 7 & -1 & 0 \\ 5 & 4 & 2 \\ -6 & 0 & 4 \end{bmatrix}.$$

2. Give the system of linear equations which is equivalent to the matrix equation

$$\begin{bmatrix} 3 & -6 \\ 2 & 1 \end{bmatrix}\begin{bmatrix} x \\ y \end{bmatrix} = \begin{bmatrix} 5 \\ 0 \end{bmatrix}.$$

3. Give a matrix equation equivalent to this system of equations:

$$\begin{cases} 8x + 3y = 7 \\ 9x - 2y = -5. \end{cases}$$

EXERCISES 3

In Exercises 1–6, give the size and special characteristics of the given matrix (such as square, column, row, identity).

1. $\begin{bmatrix} 3 & 2 & .4 \\ \frac{1}{2} & 0 & 6 \end{bmatrix}$

2. $\begin{bmatrix} 3 \\ -1 \end{bmatrix}$

3. $\begin{bmatrix} 2 & \frac{1}{3} & 0 \end{bmatrix}$

4. $\begin{bmatrix} 1 & 0 \\ 0 & 1 \end{bmatrix}$

5. $\begin{bmatrix} -2 \end{bmatrix}$

6. $\begin{bmatrix} 0 & 0 & 0 & 0 \\ 0 & 0 & 0 & 0 \end{bmatrix}$

In Exercises 7–14, perform the indicated matrix calculations.

7. $\begin{bmatrix} 4 & -2 \\ 3 & 0 \end{bmatrix} + \begin{bmatrix} 5 & 5 \\ 4 & -1 \end{bmatrix}$

8. $\begin{bmatrix} 8 \\ -3 \end{bmatrix} + \begin{bmatrix} 5 \\ 6 \end{bmatrix}$

9. $\begin{bmatrix} 2 & 8 \\ \frac{4}{3} & 4 \\ 1 & -2 \end{bmatrix} - \begin{bmatrix} 1 & 5 \\ \frac{1}{3} & 2 \\ -3 & 0 \end{bmatrix}$

10. $\begin{bmatrix} 1 & 0 \\ 0 & 1 \end{bmatrix} - \begin{bmatrix} .8 & .5 \\ .2 & .5 \end{bmatrix}$

11. $\begin{bmatrix} 5 & 3 \end{bmatrix}\begin{bmatrix} 1 \\ 2 \end{bmatrix}$

12. $\begin{bmatrix} 1 & 0 & 0 \end{bmatrix}\begin{bmatrix} \frac{1}{2} \\ 6 \\ 2 \end{bmatrix}$

13. $\begin{bmatrix} 6 & 1 & 5 \end{bmatrix}\begin{bmatrix} \frac{1}{2} \\ -3 \\ 2 \end{bmatrix}$

14. $\begin{bmatrix} 0 & 0 \end{bmatrix}\begin{bmatrix} 5 \\ -3 \end{bmatrix}$

In Exercises 15–20, the sizes of two matrices are given. Tell whether or not the product AB is defined. If so, give its size.

15. A, 3×4; B, 4×5 **16.** A, 3×3; B, 3×4 **17.** A, 3×2; B, 3×2

18. A, 1×1; B, 1×1 **19.** A, 3×3; B, 3×1 **20.** A, 4×2; B, 3×4

In Exercises 21–30, perform the multiplication.

21. $\begin{bmatrix} 3 & 1 \\ 0 & 2 \end{bmatrix}\begin{bmatrix} 1 & 4 \\ 3 & 5 \end{bmatrix}$

22. $\begin{bmatrix} 4 & -1 \\ 2 & \frac{1}{2} \end{bmatrix}\begin{bmatrix} 3 \\ 2 \end{bmatrix}$

23. $\begin{bmatrix} 4 & 1 & 0 \\ -2 & 0 & 3 \\ 1 & 5 & -1 \end{bmatrix}\begin{bmatrix} 5 \\ 1 \\ 2 \end{bmatrix}$

24. $\begin{bmatrix} 0 & 0 \\ 0 & 0 \\ 0 & 0 \end{bmatrix}\begin{bmatrix} 1 & 2 \\ 3 & 4 \end{bmatrix}$

25. $\begin{bmatrix} 1 & 0 \\ 0 & 1 \end{bmatrix}\begin{bmatrix} 5 & 6 \\ 7 & 8 \end{bmatrix}$

26. $\begin{bmatrix} 1 & 2 \\ 1 & 3 \end{bmatrix}\begin{bmatrix} 3 & -2 \\ -1 & 1 \end{bmatrix}$

27. $\begin{bmatrix} .6 & .3 \\ .4 & .7 \end{bmatrix}\begin{bmatrix} .6 & .3 \\ .4 & .7 \end{bmatrix}$

28. $\begin{bmatrix} 0 & 1 & 2 \\ -1 & 4 & \frac{1}{2} \\ 1 & 3 & 0 \end{bmatrix}\begin{bmatrix} 3 & -1 & 5 \\ 0 & 2 & 2 \\ 4 & -6 & 0 \end{bmatrix}$

29. $\begin{bmatrix} 2 & -1 & 4 \\ 0 & 1 & 0 \\ \frac{1}{2} & 3 & -2 \end{bmatrix}\begin{bmatrix} 4 & 8 & 0 \\ 3 & -1 & 2 \\ 5 & 0 & 1 \end{bmatrix}$

30. $\begin{bmatrix} 1 & 0 & 0 \\ 0 & 1 & 0 \\ 0 & 0 & 1 \end{bmatrix}\begin{bmatrix} 1 \\ 2 \\ 3 \end{bmatrix}$

In Exercises 31–34, give the system of linear equations that is equivalent to the matrix equation. Do not solve.

31. $\begin{bmatrix} 2 & 3 \\ 4 & 5 \end{bmatrix}\begin{bmatrix} x \\ y \end{bmatrix} = \begin{bmatrix} 6 \\ 7 \end{bmatrix}$

32. $\begin{bmatrix} -3 & 4 \\ 0 & 1 \end{bmatrix}\begin{bmatrix} x \\ y \end{bmatrix} = \begin{bmatrix} 1 \\ 1 \end{bmatrix}$

33. $\begin{bmatrix} 1 & 2 & 3 \\ 4 & 5 & 6 \\ 7 & 8 & 9 \end{bmatrix}\begin{bmatrix} x \\ y \\ z \end{bmatrix} = \begin{bmatrix} 10 \\ 11 \\ 12 \end{bmatrix}$

34. $\begin{bmatrix} 1 & 0 & 0 \\ 0 & 1 & 0 \\ 0 & 0 & 1 \end{bmatrix}\begin{bmatrix} x \\ y \\ z \end{bmatrix} = \begin{bmatrix} 1 \\ 2 \\ 3 \end{bmatrix}$

In Exercises 35–38, write the given system of linear equations in matrix form.

35. $\begin{cases} 3x + 2y = -1 \\ 7x - y = 2 \end{cases}$

36. $\begin{cases} 5x - 2y = 6 \\ -3x + 4y = 0 \end{cases}$

37. $\begin{cases} x - 2y + 3z = 5 \\ y + z = 6 \\ z = 2 \end{cases}$

38. $\begin{cases} -2x + 4y - z = 5 \\ x + 6y + 3z = -1 \\ 7x + 4z = 8 \end{cases}$

The distributive law says that $(A + B)C = AC + BC$. That is, adding A and B and then multiplying on the right by C gives the same result as first multiplying each of A and B on the right by C and then adding. In Exercises 39 and 40, verify the distributive law for the given matrices.

39. $A = \begin{bmatrix} 1 & 2 \\ 0 & 3 \end{bmatrix}$, $B = \begin{bmatrix} 3 & -2 \\ 4 & 5 \end{bmatrix}$, $C = \begin{bmatrix} 1 & 6 \\ 2 & 0 \end{bmatrix}$

40. $A = \begin{bmatrix} 1 & 0 & 0 \\ 0 & 1 & 0 \\ 0 & 0 & 1 \end{bmatrix}$, $B = \begin{bmatrix} 2 & 1 & 3 \\ 0 & 5 & -1 \\ 3 & 6 & 0 \end{bmatrix}$, $C = \begin{bmatrix} 0 \\ 3 \\ -4 \end{bmatrix}$

Two $n \times n$ matrices A and B are called *inverses* (of one another) if both products AB and BA equal I_n. Check that the pairs of matrices in Exercises 41 and 42 are inverses.

41. $\begin{bmatrix} 3 & -1 \\ -1 & \frac{1}{2} \end{bmatrix}, \begin{bmatrix} 1 & 2 \\ 2 & 6 \end{bmatrix}$

42. $\begin{bmatrix} 2 & 8 & -11 \\ -1 & -5 & 7 \\ 1 & 2 & -3 \end{bmatrix}, \begin{bmatrix} 1 & 2 & 1 \\ 4 & 5 & -3 \\ 3 & 4 & -2 \end{bmatrix}$

43. In a certain town the percentages of voters voting Democratic and Republican by various age groups is summarized by this matrix:

$$\begin{array}{r} \\ \text{Under 30} \\ \text{30–50} \\ \text{Over 50} \end{array} \begin{array}{c} \text{Dem.} \quad \text{Rep.} \\ \begin{bmatrix} .65 & .35 \\ .55 & .45 \\ .45 & .55 \end{bmatrix} \end{array} = A.$$

The population of voters in the town by age group is given by the matrix

$$B = [\underbrace{6000}_{\text{Under 30}} \quad \underbrace{8000}_{\text{30–50}} \quad \underbrace{4000}_{\text{Over 50}}].$$

Interpret the entries of the matrix product BA.

44. Refer to Exercise 43.

(a) Using the given data, which party would win and what would be the percentage of the winning vote?

(b) Suppose that the population of the town shifted toward older residents as reflected in the population matrix $B = [2000 \quad 4000 \quad 12{,}000]$. What would be the result of the election now?

45. Suppose that a contractor employs carpenters, bricklayers, and plumbers, working three shifts per day. The number of man-hours employed in each of the shifts is summarized in the matrix

$$\begin{array}{r} \\ \text{Carpenters} \\ \text{Bricklayers} \\ \text{Plumbers} \end{array} \begin{array}{c} \text{Shift} \\ 1 \quad 2 \quad 3 \\ \begin{bmatrix} 50 & 20 & 10 \\ 30 & 30 & 15 \\ 20 & 20 & 5 \end{bmatrix} \end{array}.$$

Labor in shift 1 costs \$10 per hour, in shift 2 \$15 per hour, and in shift 3 \$20 per hour.

(a) Without using matrix multiplication, compute the amount spent on labor in each of the shifts.

(b) Use matrix multiplication to compute the amount spent on each type of labor.

46. A flu epidemic hits a large city. Each resident of the city is either sick, well, or a carrier. The proportion of people in each of the categories is expressed by the matrix

$$\begin{array}{l} \\ \\ \text{Well} \\ \text{Sick} \\ \text{Carrier} \end{array} \begin{array}{c} \text{Age} \\ \hline \begin{array}{ccc} 0\text{–}10 & 10\text{–}30 & \text{Over } 30 \end{array} \\ \begin{bmatrix} .70 & .70 & .60 \\ .10 & .20 & .30 \\ .20 & .10 & .10 \end{bmatrix} \end{array} = A.$$

The population of the city is distributed by age and sex as follows:

$$\text{Age} \left| \begin{array}{l} 0\text{–}10 \\ 10\text{–}30 \\ \text{Over } 30 \end{array} \right. \begin{array}{c} \begin{array}{cc} \text{Male} & \text{Female} \end{array} \\ \begin{bmatrix} 60{,}000 & 65{,}000 \\ 100{,}000 & 110{,}000 \\ 200{,}000 & 230{,}000 \end{bmatrix} \end{array} = B.$$

(a) Compute AB.

(b) How many sick males are there?

(c) How many female carriers are there?

SOLUTIONS TO PRACTICE PROBLEMS 3

1. Answer:

$$\begin{bmatrix} 3 & 1 & 2 \\ -1 & 0 & \frac{1}{2} \\ 0 & 4 & 1 \end{bmatrix} \begin{bmatrix} 7 & -1 & 0 \\ 5 & 4 & 2 \\ -6 & 0 & 4 \end{bmatrix} = \begin{bmatrix} 14 & 1 & 10 \\ -10 & 1 & 2 \\ 14 & 16 & 12 \end{bmatrix}.$$

The systematic steps to be taken are:

(a) Determine the size of the product matrix.
Since we have a ③ × 3 times a 3 × ③, the size of the product is given by the outer dimensions or 3 × 3. Begin by drawing a 3 × 3 rectangular array.
(outer dimensions)

(b) Find the entries one at a time.
To find the entry in the first row, first column of the product, look at the first row of the left given matrix and the first column of the right given matrix and form their product.

$$\begin{bmatrix} 3 & 1 & 2 \\ -1 & 0 & \frac{1}{2} \\ 0 & 4 & 1 \end{bmatrix} \begin{bmatrix} 7 & -1 & 0 \\ 5 & 4 & 2 \\ -6 & 0 & 4 \end{bmatrix} = \begin{bmatrix} 14 & & \\ & & \\ & & \end{bmatrix},$$

since $3 \cdot 7 + 1 \cdot 5 + 2(-6) = 14$. In general, to find the entry in the ith row, jth column of the product, put one finger on the ith row of the left given matrix and another finger on the jth column of the right given matrix. Then multiply the row matrix times the column matrix to get the desired entry.

2. Denote the three matrices by A, X, and B, respectively. Since b_{11} (the entry of the first row, first column of B) is 5, this means that

$$[\text{first row of } A]\begin{bmatrix}\text{first}\\ \text{column}\\ \text{of } X\end{bmatrix} = [b_{11}].$$

That is,

$$\begin{bmatrix}3 & -6\end{bmatrix}\begin{bmatrix}x\\ y\end{bmatrix} = [5]$$

or

$$3x - 6y = 5.$$

Similarly, $b_{21} = 0$ says that $2x + y = 0$. Therefore, the corresponding system of linear equations is

$$\begin{cases}3x - 6y = 5\\ 2x + y = 0.\end{cases}$$

3. The coefficient matrix is

$$\begin{bmatrix}8 & 3\\ 9 & -2\end{bmatrix}.$$

So the system of equations is equivalent to the matrix equation

$$\begin{bmatrix}8 & 3\\ 9 & -2\end{bmatrix}\begin{bmatrix}x\\ y\end{bmatrix} = \begin{bmatrix}7\\ -5\end{bmatrix}.$$

1.4 The Inverse of a Matrix

In the preceding section we introduced the operations of addition, subtraction, and multiplication of matrices. In this section let us pursue the algebra of matrices a bit further and consider equations involving matrices. Specifically, we shall consider equations of the form

$$AX = B, \tag{1}$$

where A and B are given matrices and X is an unknown matrix whose entries are to be determined. Such equations among matrices are intimately bound up with the theory of systems of linear equations. Indeed, we described the connection in a special case in Example 6 of the preceding section. In that example we wrote the system of linear equations

$$\begin{cases}-2x + 4y = 2\\ -3x + 7y = 7\end{cases}$$

as a matrix equation of the form (1), where

$$A = \begin{bmatrix}-2 & 4\\ -3 & 7\end{bmatrix}, \quad B = \begin{bmatrix}2\\ 7\end{bmatrix}, \quad X = \begin{bmatrix}x\\ y\end{bmatrix}.$$

Note that by determining the entries (x and y) of the unknown matrix X, we solve the system of linear equations. We will return to this example after we have made a complete study of the matrix equation (1).

As motivation for our solution of equation (1), let us consider the analogous equation among numbers:

$$ax = b,$$

where a and b are given numbers* and x is to be determined. Let us examine its solution in great detail. Multiply both sides by $1/a$. (Note that $1/a$ makes sense, since $a \neq 0$.)

$$\left(\frac{1}{a}\right) \cdot (ax) = \frac{1}{a} \cdot b$$

$$\left(\frac{1}{a} \cdot a\right) \cdot x = \frac{1}{a} \cdot b$$

$$1 \cdot x = \frac{1}{a} \cdot b$$

$$x = \frac{1}{a} \cdot b.$$

Let us model our solution of equation (1) on the calculation above. To do so, we wish to multiply both sides of the equation by a matrix that plays the same role in matrix arithmetic as $1/a$ plays in ordinary arithmetic. Our first task then will be to introduce this matrix and study its properties.

The number $1/a$ has the following relationship to the number a:

$$\frac{1}{a} \cdot a = a \cdot \frac{1}{a} = 1. \tag{2}$$

The matrix analog of the number 1 is an identity matrix I. This prompts us to generalize equation (2) to matrices as follows. Suppose that we are given a square matrix A. Then the *inverse* of A, denoted A^{-1}, is a square matrix with the property

$$A^{-1}A = I \quad \text{and} \quad AA^{-1} = I, \tag{3}$$

where I is an identity matrix of the same size as A. The matrix A^{-1} is the matrix analogue of the number $1/a$. It can be shown that a matrix A has at most one inverse. (However, A may not have an inverse at all; see below.)

If we are given a matrix A, then it is easy to determine whether or not a given matrix is its inverse. Merely check equation (3) with the given matrix substituted for A^{-1}. For example, if

$$A = \begin{bmatrix} -2 & 4 \\ -3 & 7 \end{bmatrix},$$

---

* We may as well assume that $a \neq 0$. Otherwise, x does not occur.

then

$$A^{-1} = \begin{bmatrix} -\frac{7}{2} & 2 \\ -\frac{3}{2} & 1 \end{bmatrix}.$$

Indeed, we have

$$\underbrace{\begin{bmatrix} -\frac{7}{2} & 2 \\ -\frac{3}{2} & 1 \end{bmatrix}}_{A^{-1}} \underbrace{\begin{bmatrix} -2 & 4 \\ -3 & 7 \end{bmatrix}}_{A} = \begin{bmatrix} 7 - 6 & -14 + 14 \\ 3 - 3 & -6 + 7 \end{bmatrix} = \underbrace{\begin{bmatrix} 1 & 0 \\ 0 & 1 \end{bmatrix}}_{I_2}$$

and

$$\underbrace{\begin{bmatrix} -2 & 4 \\ -3 & 7 \end{bmatrix}}_{A} \underbrace{\begin{bmatrix} -\frac{7}{2} & 2 \\ -\frac{3}{2} & 1 \end{bmatrix}}_{A^{-1}} = \begin{bmatrix} 7 - 6 & -4 + 4 \\ \frac{21}{2} - \frac{21}{2} & -6 + 7 \end{bmatrix} = \underbrace{\begin{bmatrix} 1 & 0 \\ 0 & 1 \end{bmatrix}}_{I_2}.$$

The inverse of a matrix can be calculated using Gaussian elimination, as the next example illustrates.

EXAMPLE 1 Let $A = \begin{bmatrix} 3 & 1 \\ 5 & 2 \end{bmatrix}$. Determine A^{-1}.

Solution Since A is a 2×2 matrix, A^{-1} is also a 2×2 matrix and satisfies

$$AA^{-1} = I_2 \qquad \text{and} \qquad A^{-1}A = I_2, \tag{4}$$

where $I_2 = \begin{bmatrix} 1 & 0 \\ 0 & 1 \end{bmatrix}$ is a 2×2 identity matrix. Suppose that

$$A^{-1} = \begin{bmatrix} x & y \\ z & w \end{bmatrix}.$$

Then the first equation of (4) reads

$$\begin{bmatrix} 3 & 1 \\ 5 & 2 \end{bmatrix} \begin{bmatrix} x & y \\ z & w \end{bmatrix} = \begin{bmatrix} 1 & 0 \\ 0 & 1 \end{bmatrix}.$$

Multiplying out the matrices on the left gives

$$\begin{bmatrix} 3x + z & 3y + w \\ 5x + 2z & 5y + 2w \end{bmatrix} = \begin{bmatrix} 1 & 0 \\ 0 & 1 \end{bmatrix}.$$

Now equate corresponding elements in the two matrices to obtain the equations

$$\begin{cases} 3x + z = 1 \\ 5x + 2z = 0, \end{cases} \qquad \begin{cases} 3y + w = 0 \\ 5y + 2w = 1. \end{cases}$$

Notice that the equations break up into two pairs of linear equations, each pair involving only two variables. Solving these two systems of linear equations yields $x = 2, z = -5, y = -1, w = 3$. Therefore,

$$A^{-1} = \begin{bmatrix} 2 & -1 \\ -5 & 3 \end{bmatrix}.$$

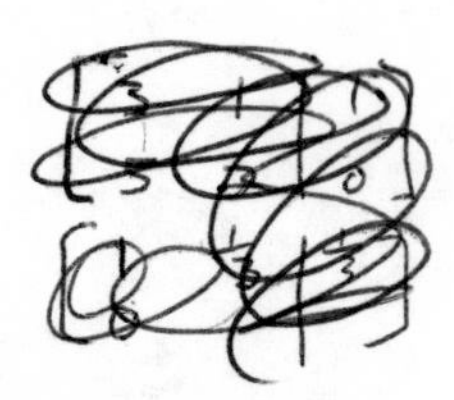

Indeed, we may readily verify that

$$\begin{bmatrix} 3 & 1 \\ 5 & 2 \end{bmatrix}\begin{bmatrix} 2 & -1 \\ -5 & 3 \end{bmatrix} = \begin{bmatrix} 1 & 0 \\ 0 & 1 \end{bmatrix}$$

$$\begin{bmatrix} 2 & -1 \\ -5 & 3 \end{bmatrix}\begin{bmatrix} 3 & 1 \\ 5 & 2 \end{bmatrix} = \begin{bmatrix} 1 & 0 \\ 0 & 1 \end{bmatrix}.$$

The method above can be used to calculate the inverse of matrices of any size, although it involves considerable calculation. We shall provide a rather efficient computational method for calculating A^{-1} in the next section. For now, however, let us be content with the above method. Using it, we can derive a general formula for A^{-1} in case A is a 2×2 matrix.

To determine the inverse of a 2×2 matrix Let

$$A = \begin{bmatrix} a & b \\ c & d \end{bmatrix}.$$

Let $\Delta = ad - bc$ and assume that $\Delta \neq 0$. Then A^{-1} is given by the formula

$$A^{-1} = \begin{bmatrix} \frac{d}{\Delta} & -\frac{b}{\Delta} \\ -\frac{c}{\Delta} & \frac{a}{\Delta} \end{bmatrix}. \tag{5}$$

We will omit the derivation of this formula. It proceeds along lines similar to those of Example 1. Notice that formula (5) involves division by Δ. Since division by 0 is not permissible, it is necessary that $\Delta \neq 0$ for formula (5) to be applied. We will discuss the case $\Delta = 0$ below.

Equation (5) can be reduced to a simple step-by-step procedure.

To determine the inverse of $\begin{bmatrix} a & b \\ c & d \end{bmatrix}$ *if* $\Delta = ad - bc \neq 0$:

1. Interchange a and d to get $\begin{bmatrix} d & b \\ c & a \end{bmatrix}$.
2. Change the signs of b and c to get $\begin{bmatrix} d & -b \\ -c & a \end{bmatrix}$.
3. Divide all entries by Δ to get $\begin{bmatrix} \frac{d}{\Delta} & -\frac{b}{\Delta} \\ -\frac{c}{\Delta} & \frac{a}{\Delta} \end{bmatrix}$.

EXAMPLE 2 Calculate the inverse of $\begin{bmatrix} -2 & 4 \\ -3 & 7 \end{bmatrix}$.

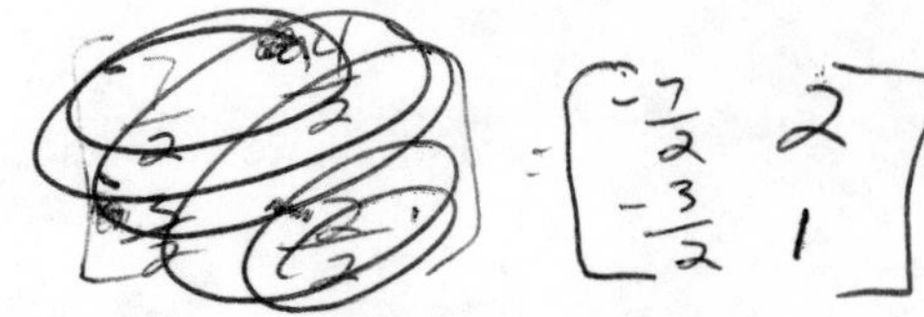

Solution $\Delta = (-2) \cdot 7 - 4 \cdot (-3) = -2$, so $\Delta \neq 0$, and we may use the computation above.

1. Interchange a and d:

$$\begin{bmatrix} 7 & 4 \\ -3 & -2 \end{bmatrix}$$

2. Change signs of b and c:

$$\begin{bmatrix} 7 & -4 \\ 3 & -2 \end{bmatrix}$$

3. Divide all entries by $\Delta = -2$:

$$\begin{bmatrix} -\frac{7}{2} & 2 \\ -\frac{3}{2} & 1 \end{bmatrix}$$

Thus

$$\begin{bmatrix} -2 & 4 \\ -3 & 7 \end{bmatrix}^{-1} = \begin{bmatrix} -\frac{7}{2} & 2 \\ -\frac{3}{2} & 1 \end{bmatrix}.$$

Not every square matrix has an inverse. Indeed, it may be impossible to satisfy equations (3) for any choice of A^{-1}. This phenomenon can even occur in the case of 2×2 matrices. Here, one can show that *if* $\Delta = 0$, *then the matrix does not have an inverse.* The next example illustrates this phenomenon in a special case.

EXAMPLE 3 Show that $\begin{bmatrix} 1 & 1 \\ 1 & 1 \end{bmatrix}$ does not have an inverse.

Solution Note first that $\Delta = 1 \cdot 1 - 1 \cdot 1 = 0$, so the inverse cannot be computed via equation (5). Suppose that the given matrix did have an inverse, say

$$\begin{bmatrix} s & t \\ u & v \end{bmatrix}.$$

Then the following equation would hold:

$$\begin{bmatrix} s & t \\ u & v \end{bmatrix} \begin{bmatrix} 1 & 1 \\ 1 & 1 \end{bmatrix} = \begin{bmatrix} 1 & 0 \\ 0 & 1 \end{bmatrix}.$$

On multiplying out the two matrices on the left, we get the equation

$$\begin{bmatrix} s + t & s + t \\ u + v & u + v \end{bmatrix} = \begin{bmatrix} 1 & 0 \\ 0 & 1 \end{bmatrix},$$

or, on equating entries in the first row:

$$s + t = 1, \qquad s + t = 0.$$

But $s + t$ cannot equal both 1 and 0. So we reach a contradiction, and therefore the original matrix cannot have an inverse.

We were led to introduce the inverse of a matrix from a discussion of the matrix equation $AX = B$. Let us now return to that discussion. Suppose that A and B are given matrices and that we wish to solve the matrix equation

$$AX = B$$

for the unknown matrix X. Suppose further that A has an inverse A^{-1}. Multiply both sides of the equation on the left by A^{-1} to obtain

$$A^{-1} \cdot AX = A^{-1}B.$$

Because $A^{-1} \cdot A = I$, we have

$$IX = A^{-1}B$$

$$X = A^{-1}B.$$

Thus the matrix X is found by simply multiplying B on the left by A^{-1}, and we can summarize our findings as follows:

Solving a Matrix Equation If the matrix A has an inverse, then the solution of the matrix equation

$$AX = B$$

is given by

$$X = A^{-1}B.$$

Matrix equations can be used to solve systems of linear equations, as illustrated in the next example.

EXAMPLE 4 Use a matrix equation to solve the system of linear equations

$$\begin{cases} -2x + 4y = 2 \\ -3x + 7y = 7. \end{cases}$$

Solution In Example 6 of the preceding section we saw that the system could be written as a matrix equation:

$$\underset{A}{\begin{bmatrix} -2 & 4 \\ -3 & 7 \end{bmatrix}} \underset{X}{\begin{bmatrix} x \\ y \end{bmatrix}} = \underset{B}{\begin{bmatrix} 2 \\ 7 \end{bmatrix}}.$$

We happen to know A^{-1} from Example 2, namely

$$A^{-1} = \begin{bmatrix} -\frac{7}{2} & 2 \\ -\frac{3}{2} & 1 \end{bmatrix}.$$

So we may compute the matrix $X = A^{-1}B$:

$$X = \begin{bmatrix} x \\ y \end{bmatrix} = \begin{bmatrix} -\frac{7}{2} & 2 \\ -\frac{3}{2} & 1 \end{bmatrix}\begin{bmatrix} 2 \\ 7 \end{bmatrix} = \begin{bmatrix} 7 \\ 4 \end{bmatrix}.$$

Thus the solution of the system is $x = 7$, $y = 4$.

Here is an application of matrix questions which is a preview of the discussion of stochastic matrices in Chapter 8.*

EXAMPLE 5 Let x and y denote the number of married and single adults in a certain town as of January 1. Let m and s denote the corresponding numbers for the following year. A statistical survey shows that x, y, m, and s are related by the equations

$$.9x + .2y = m$$

$$.1x + .8y = s.$$

In a given year there were found to be 490,000 married adults and 147,000 single adults.

(a) How many married adults were there in the preceding year?

(b) How many married adults were there 2 years ago?

Solution (a) The given equations can be written in the matrix form

$$AX = B,$$

where

$$A = \begin{bmatrix} .9 & .2 \\ .1 & .8 \end{bmatrix}, \quad X = \begin{bmatrix} x \\ y \end{bmatrix}, \quad B = \begin{bmatrix} m \\ s \end{bmatrix}.$$

We are given that $B = \begin{bmatrix} 490{,}000 \\ 147{,}000 \end{bmatrix}$. So, since

$$X = A^{-1}B$$

and

$$A^{-1} = \begin{bmatrix} \frac{8}{7} & -\frac{2}{7} \\ -\frac{1}{7} & \frac{9}{7} \end{bmatrix},$$

we have

$$X = \begin{bmatrix} \frac{8}{7} & -\frac{2}{7} \\ -\frac{1}{7} & \frac{9}{7} \end{bmatrix}\begin{bmatrix} 490{,}000 \\ 147{,}000 \end{bmatrix} = \begin{bmatrix} 518{,}000 \\ 119{,}000 \end{bmatrix}.$$

* The reader is referred to *Mathematics for the Management, Life, and Social Sciences by* Goldstein, Lay, and Schneider

Thus last year there were 518,000 married adults and 119,000 single adults.

(b) We deduce x and y for two years ago from the values of m and s for last year, namely $m = 518{,}000$, $s = 119{,}000$.

$$X = A^{-1}B = \begin{bmatrix} \frac{8}{7} & -\frac{2}{7} \\ -\frac{1}{7} & \frac{9}{7} \end{bmatrix}\begin{bmatrix} 518{,}000 \\ 119{,}000 \end{bmatrix} = \begin{bmatrix} 558{,}000 \\ 79{,}000 \end{bmatrix}.$$

That is, 2 years ago there were 558,000 married adults and 79,000 single adults.

EXAMPLE 6 In the next section we will show that if

$$A = \begin{bmatrix} 4 & -2 & 3 \\ 8 & -3 & 5 \\ 7 & -2 & 4 \end{bmatrix}, \quad \text{then} \quad A^{-1} = \begin{bmatrix} -2 & 2 & -1 \\ 3 & -5 & 4 \\ 5 & -6 & 4 \end{bmatrix}.$$

(a) Use this fact to solve the system of linear equations

$$\begin{cases} 4x - 2y + 3z = 1 \\ 8x - 3y + 5z = 4 \\ 7x - 2y + 4z = 5. \end{cases}$$

(b) Solve the system of equations

$$\begin{cases} 4x - 2y + 3z = 4 \\ 8x - 3y + 5z = 7 \\ 7x - 2y + 4z = 6. \end{cases}$$

Solution (a) The system can be written in the matrix form

$$\underset{A}{\begin{bmatrix} 4 & -2 & 3 \\ 8 & -3 & 5 \\ 7 & -2 & 4 \end{bmatrix}} \underset{X}{\begin{bmatrix} x \\ y \\ z \end{bmatrix}} = \underset{B}{\begin{bmatrix} 1 \\ 4 \\ 5 \end{bmatrix}}.$$

The solution of this matrix equation is $X = A^{-1}B$ or

$$\begin{bmatrix} x \\ y \\ z \end{bmatrix} = \begin{bmatrix} -2 & 2 & -1 \\ 3 & -5 & 4 \\ 5 & -6 & 4 \end{bmatrix}\begin{bmatrix} 1 \\ 4 \\ 5 \end{bmatrix} = \begin{bmatrix} 1 \\ 3 \\ 1 \end{bmatrix}.$$

Thus the solution of the system is $x = 1$, $y = 3$, $z = 1$.

(b) This system has the same left-hand side as the preceding system, so its solution is

$$\begin{bmatrix} x \\ y \\ z \end{bmatrix} = \begin{bmatrix} -2 & 2 & -1 \\ 3 & -5 & 4 \\ 5 & -6 & 4 \end{bmatrix}\begin{bmatrix} 4 \\ 7 \\ 6 \end{bmatrix} = \begin{bmatrix} 0 \\ 1 \\ 2 \end{bmatrix}.$$

That is, the solution of the system is $x = 0$, $y = 1$, $z = 2$.

Using the method of matrix equations to solve a system of linear equations is especially efficient if one wishes to solve a number of systems all having the same left-hand sides but different right-hand sides. For then A^{-1} must be computed only once for all the systems under consideration. (This point is useful in Exercises 17–20.)

PRACTICE PROBLEMS 4

1. Show that the inverse of

$$\begin{bmatrix} -4 & 1 & 2 \\ 7 & -1 & -4 \\ -\frac{1}{2} & 0 & \frac{1}{2} \end{bmatrix} \quad \text{is} \quad \begin{bmatrix} 1 & 1 & 4 \\ 3 & 2 & 4 \\ 1 & 1 & 6 \end{bmatrix}.$$

2. Use the method of this section to solve the system of linear equations

$$\begin{cases} .8x + .6y = 5 \\ .2x + .4y = 2. \end{cases}$$

EXERCISES 4

In Exercises 1 and 2, use the fact that

$$\begin{bmatrix} 2 & 2 \\ \frac{1}{2} & 1 \end{bmatrix}^{-1} = \begin{bmatrix} 1 & -2 \\ -\frac{1}{2} & 2 \end{bmatrix}.$$

1. Solve $\begin{cases} 2x + 2y = 4 \\ \frac{1}{2}x + \ \ y = 1. \end{cases}$

2. Solve $\begin{cases} 2x + 2y = 14 \\ \frac{1}{2}x + \ \ y = 4. \end{cases}$

In Exercises 3–10, find the inverse of the given matrix.

3. $\begin{bmatrix} 7 & 2 \\ 3 & 1 \end{bmatrix}$

4. $\begin{bmatrix} 2 & 3 \\ 5 & 7 \end{bmatrix}$

5. $\begin{bmatrix} 6 & 2 \\ 5 & 2 \end{bmatrix}$

6. $\begin{bmatrix} 1 & .5 \\ 0 & .5 \end{bmatrix}$

7. $\begin{bmatrix} .7 & .2 \\ .3 & .8 \end{bmatrix}$

8. $\begin{bmatrix} 0 & 1 \\ 1 & 0 \end{bmatrix}$

9. $[3]$

10. $[.2]$

In Exercises 11–14, use the method of this section to solve the system of linear equations.

11. $\begin{cases} x + 2y = 3 \\ 2x + 6y = 5 \end{cases}$

12. $\begin{cases} 5x + 3y = 1 \\ 7x + 4y = 2 \end{cases}$

13. $\begin{cases} \frac{1}{2}x + \ 2y = 4 \\ 3x + 16y = 0 \end{cases}$

14. $\begin{cases} .8x + .6y = 2 \\ .2x + .4y = 1 \end{cases}$

15. It is found that the number of married and single adults in a certain town are subject to the following statistics. Suppose that x and y denote the number of married and single adults, respectively, in a given year (say as of January 1) and let m, s denote the corre-

sponding numbers for the following year. Then

$$.8x + .3y = m$$
$$.2x + .7y = s.$$

(a) Write this system of equations in matrix form.

(b) Solve the resulting matrix equation for $X = \begin{bmatrix} x \\ y \end{bmatrix}$.

(c) Suppose that in a given year there were found to be 100,000 married adults and 50,000 single adults. How many married (resp. single) adults were there the preceding year?

(d) How many married (resp. single) adults were there 2 years ago?

16. A flu epidemic is spreading through a town of 48,000 people. It is found that if x and y denote the numbers of people sick and well in a given week, respectively, and if s and w denote the corresponding numbers for the following week, then

$$\tfrac{1}{3}x + \tfrac{1}{4}y = s$$
$$\tfrac{2}{3}x + \tfrac{3}{4}y = w.$$

(a) Write this system of equations in matrix form.

(b) Solve the resulting matrix equation for $X = \begin{bmatrix} x \\ y \end{bmatrix}$.

(c) Suppose that 13,000 people are sick in a given week. How many were sick the preceding week?

(d) Same question as part (c), except assume that 14,000 are sick.

In Exercises 17 and 18, use the fact that

$$\begin{bmatrix} 1 & 2 & 2 \\ 1 & 3 & 2 \\ 1 & 2 & 3 \end{bmatrix}^{-1} = \begin{bmatrix} 5 & -2 & -2 \\ -1 & 1 & 0 \\ -1 & 0 & 1 \end{bmatrix}.$$

17. Solve $\begin{cases} x + 2y + 2z = 1 \\ x + 3y + 2z = -1 \\ x + 2y + 3z = -1. \end{cases}$

18. Solve $\begin{cases} x + 2y + 2z = 1 \\ x + 3y + 2z = 0 \\ x + 2y + 3z = 0. \end{cases}$

In Exercises 19 and 20, use the fact that

$$\begin{bmatrix} 9 & 0 & 2 & 0 \\ -20 & -9 & -5 & 5 \\ 4 & 0 & 1 & 0 \\ -4 & -2 & -1 & 1 \end{bmatrix}^{-1} = \begin{bmatrix} 1 & 0 & -2 & 0 \\ 0 & 1 & 0 & -5 \\ -4 & 0 & 9 & 0 \\ 0 & 2 & 1 & -9 \end{bmatrix}.$$

19. Solve $\begin{cases} 9x + 2z = 1 \\ -20x - 9y - 5z + 5w = 0 \\ 4x + z = 0 \\ -4x - 2y - z + w = -1. \end{cases}$

20. Solve $\begin{cases} 9x + 2z = 2 \\ -20x - 9y - 5z + 5w = 1 \\ 4x + z = 3 \\ -4x - 2y - z + w = 0. \end{cases}$

SOLUTIONS TO PRACTICE PROBLEMS 4

1. To see if this matrix is indeed the inverse, multiply it by the original matrix and find out if the products are identity matrices.

$$\begin{bmatrix} 1 & 1 & 4 \\ 3 & 2 & 4 \\ 1 & 1 & 6 \end{bmatrix}\begin{bmatrix} -4 & 1 & 2 \\ 7 & -1 & -4 \\ -\frac{1}{2} & 0 & \frac{1}{2} \end{bmatrix} = \begin{bmatrix} 1 & 0 & 0 \\ 0 & 1 & 0 \\ 0 & 0 & 1 \end{bmatrix}, \quad \text{an identity matrix.}$$

$$\begin{bmatrix} -4 & 1 & 2 \\ 7 & -1 & -4 \\ -\frac{1}{2} & 0 & \frac{1}{2} \end{bmatrix}\begin{bmatrix} 1 & 1 & 4 \\ 3 & 2 & 4 \\ 1 & 1 & 6 \end{bmatrix} = \begin{bmatrix} 1 & 0 & 0 \\ 0 & 1 & 0 \\ 0 & 0 & 1 \end{bmatrix}.$$

2. The matrix form of this system is

$$\begin{bmatrix} .8 & .6 \\ .2 & .4 \end{bmatrix}\begin{bmatrix} x \\ y \end{bmatrix} = \begin{bmatrix} 5 \\ 2 \end{bmatrix}.$$

Therefore, the solution is

$$\begin{bmatrix} x \\ y \end{bmatrix} = \begin{bmatrix} .8 & .6 \\ .2 & .4 \end{bmatrix}^{-1}\begin{bmatrix} 5 \\ 2 \end{bmatrix}.$$

To compute the inverse of the 2×2 matrix, first compute Δ.

$$\Delta = ad - bc = (.8)(.4) - (.6)(.2) = .32 - .12 = .2.$$

Thus

$$\begin{bmatrix} .8 & .6 \\ .2 & .4 \end{bmatrix}^{-1} = \begin{bmatrix} .4/.2 & -.6/.2 \\ -.2/.2 & .8/.2 \end{bmatrix} = \begin{bmatrix} 2 & -3 \\ -1 & 4 \end{bmatrix}.$$

Therefore,

$$\begin{bmatrix} x \\ y \end{bmatrix} = \begin{bmatrix} 2 & -3 \\ -1 & 4 \end{bmatrix}\begin{bmatrix} 5 \\ 2 \end{bmatrix} = \begin{bmatrix} 4 \\ 3 \end{bmatrix}.$$

So the solution is $x = 4$, $y = 3$.

1.5 The Gauss-Jordan Method for Calculating Inverses

Of the several popular methods for finding the inverse of a matrix, the Gauss-Jordan method is probably the easiest to describe. It can be used on square matrices of any size. Also, the mechanical nature of the computations allows this method to be programmed for a computer with relative ease. We shall illustrate the procedure with a 2×2 matrix, whose inverse can also be calculated using the method of the previous section. Let

$$A = \begin{bmatrix} \frac{1}{2} & 1 \\ 1 & 3 \end{bmatrix}.$$

It is simple to check that

$$A^{-1} = \begin{bmatrix} 6 & -2 \\ -2 & 1 \end{bmatrix}.$$

Let us now derive this result using the Gauss-Jordan method.

Step 1 Write down the matrix A, and on its right an identity matrix of the same size.

This is most conveniently done by placing I_2 beside A in a single matrix.

$$\left[\underbrace{\begin{matrix} \frac{1}{2} & 1 \\ 1 & 3 \end{matrix}}_{A} \middle| \underbrace{\begin{matrix} 1 & 0 \\ 0 & 1 \end{matrix}}_{I_2}\right].$$

Step 2 Perform elementary row operations on the left-hand matrix so as to transform it into an identity matrix. Each operation performed on the left-hand matrix is also performed on the right-hand matrix.

This step proceeds exactly like the Gaussian elimination method and may be most conveniently expressed in terms of pivoting.

$$\left[\begin{matrix} \textcircled{\tfrac{1}{2}} & 1 \\ 1 & 3 \end{matrix} \middle| \begin{matrix} 1 & 0 \\ 0 & 1 \end{matrix}\right], \quad \left[\begin{matrix} 1 & 2 \\ 0 & \textcircled{1} \end{matrix} \middle| \begin{matrix} 2 & 0 \\ -2 & 1 \end{matrix}\right], \quad \left[\begin{matrix} 1 & 0 \\ 0 & 1 \end{matrix} \middle| \begin{matrix} 6 & -2 \\ -2 & 1 \end{matrix}\right].$$

Step 3 When the matrix on the left becomes an identity matrix, the matrix on the right is the desired inverse.

So, from the last matrix of our calculation above, we have

$$A^{-1} = \begin{bmatrix} 6 & -2 \\ -2 & 1 \end{bmatrix}.$$

This is the same result obtained earlier.

We will demonstrate why the method above works after some further examples.

EXAMPLE 1 Find the inverse of the matrix

$$A = \begin{bmatrix} 4 & -2 & 3 \\ 8 & -3 & 5 \\ 7 & -2 & 4 \end{bmatrix}.$$

Solution

$$\left[\begin{array}{ccc|ccc} \textcircled{4} & -2 & 3 & 1 & 0 & 0 \\ 8 & -3 & 5 & 0 & 1 & 0 \\ 7 & -2 & 4 & 0 & 0 & 1 \end{array}\right]$$

$$\left[\begin{array}{ccc|ccc} 1 & -\frac{1}{2} & \frac{3}{4} & \frac{1}{4} & 0 & 0 \\ 0 & \textcircled{1} & -1 & -2 & 1 & 0 \\ 0 & \frac{3}{2} & -\frac{5}{4} & -\frac{7}{4} & 0 & 1 \end{array}\right]$$

$$\left[\begin{array}{ccc|ccc} 1 & 0 & \frac{1}{4} & -\frac{3}{4} & \frac{1}{2} & 0 \\ 0 & 1 & -1 & -2 & 1 & 0 \\ 0 & 0 & \textcircled{$\frac{1}{4}$} & \frac{5}{4} & -\frac{3}{2} & 1 \end{array}\right]$$

$$\left[\begin{array}{ccc|ccc} 1 & 0 & 0 & -2 & 2 & -1 \\ 0 & 1 & 0 & 3 & -5 & 4 \\ 0 & 0 & 1 & 5 & -6 & 4 \end{array}\right].$$

Therefore,

$$A^{-1} = \begin{bmatrix} -2 & 2 & -1 \\ 3 & -5 & 4 \\ 5 & -6 & 4 \end{bmatrix}.$$

Not all square matrices have inverses. If a matrix does not have an inverse, this will become apparent when applying the Gauss-Jordan method. At some point there will be no way to continue transforming the left-hand matrix into an identity matrix. This is illustrated in the next example.

EXAMPLE 2 Find the inverse of the matrix

$$A = \begin{bmatrix} 1 & 3 & 2 \\ 0 & 1 & 4 \\ 1 & 5 & 10 \end{bmatrix}.$$

Solution

$$\left[\begin{array}{ccc|ccc} 1 & 3 & 2 & 1 & 0 & 0 \\ \textcircled{0} & 1 & 4 & 0 & 1 & 0 \\ 1 & 5 & 10 & 0 & 0 & 1 \end{array}\right]$$

$$\left[\begin{array}{ccc|ccc} 1 & 3 & 2 & 1 & 0 & 0 \\ 0 & \textcircled{1} & 4 & 0 & 1 & 0 \\ 0 & 2 & 8 & -1 & 0 & 1 \end{array}\right]$$

$$\left[\begin{array}{ccc|ccc} 1 & 0 & -10 & 1 & -3 & 0 \\ 0 & 1 & 4 & 0 & 1 & 0 \\ 0 & 0 & 0 & -1 & -2 & 1 \end{array}\right].$$

Since the third row of the left-hand matrix has only zero entries, it is impossible to complete the Gauss-Jordan method. Therefore, the matrix A has no inverse matrix.

Verification of the Gauss-Jordan Method for Calculating Inverses

In the preceding section we showed how to calculate the inverse by solving several systems of linear equations. Actually, the Gauss-Jordan method is just an organized way of going about the calculation. To see why, let us consider a concrete example:

$$A = \begin{bmatrix} 4 & -2 & 3 \\ 8 & -3 & 5 \\ 7 & -2 & 4 \end{bmatrix}.$$

We wish to determine A^{-1}, so regard it as a matrix of unknowns:

$$A^{-1} = \begin{bmatrix} x_1 & x_2 & x_3 \\ y_1 & y_2 & y_3 \\ z_1 & z_2 & z_3 \end{bmatrix}.$$

The statement $AA^{-1} = I_3$ is

$$\begin{bmatrix} 4 & -2 & 3 \\ 8 & -3 & 5 \\ 7 & -2 & 4 \end{bmatrix} \begin{bmatrix} x_1 & x_2 & x_3 \\ y_1 & y_2 & y_3 \\ z_1 & z_2 & z_3 \end{bmatrix} = \begin{bmatrix} 1 & 0 & 0 \\ 0 & 1 & 0 \\ 0 & 0 & 1 \end{bmatrix}.$$

Multiplying out the matrices on the left and comparing the result with the matrix on the right gives us nine equations, namely

$$\begin{cases} 4x_1 - 2y_1 + 3z_1 = 1 \\ 8x_1 - 3y_1 + 5z_1 = 0 \\ 7x_1 - 2y_1 + 4z_1 = 0 \end{cases}$$

$$\begin{cases} 4x_2 - 2y_2 + 3z_2 = 0 \\ 8x_2 - 3y_2 + 5z_2 = 1 \\ 7x_2 - 2y_2 + 4z_2 = 0 \end{cases}$$

$$\begin{cases} 4x_3 - 2y_3 + 3z_3 = 0 \\ 8x_3 - 3y_3 + 5z_3 = 0 \\ 7x_3 - 2y_3 + 4z_3 = 1. \end{cases}$$

Notice that each system of equations corresponds to one column of unknowns in A^{-1}. More precisely, if we set

$$X_1 = \begin{bmatrix} x_1 \\ y_1 \\ z_1 \end{bmatrix}, \quad X_2 = \begin{bmatrix} x_2 \\ y_2 \\ z_2 \end{bmatrix}, \quad X_3 = \begin{bmatrix} x_3 \\ y_3 \\ z_3 \end{bmatrix},$$

then the three systems above have the respective matrix forms

$$AX_1 = \begin{bmatrix} 1 \\ 0 \\ 0 \end{bmatrix}, \quad AX_2 = \begin{bmatrix} 0 \\ 1 \\ 0 \end{bmatrix}, \quad AX_3 = \begin{bmatrix} 0 \\ 0 \\ 1 \end{bmatrix}.$$

Now imagine the process of applying Gaussian elimination to solve these three systems. We apply elementary row operations to the matrices

$$\left[\begin{array}{c|c} & 1 \\ A & 0 \\ & 0 \end{array}\right], \quad \left[\begin{array}{c|c} & 0 \\ A & 1 \\ & 0 \end{array}\right], \quad \left[\begin{array}{c|c} & 0 \\ A & 0 \\ & 1 \end{array}\right].$$

The process ends when we convert A into the identity matrix, at which point the solutions may be read off the right column. So the procedure ends with the matrices

$$[I_3 \mid X_1], \quad [I_3 \mid X_2], \quad [I_3 \mid X_3].$$

Realize, however, that at each step of the three Gaussian eliminations we are performing the same operations, since all three start with the matrix A on the left. So, in order to save calculation, perform the three Gaussian eliminations simultaneously by performing the row operations on the composite matrix

$$\left[\begin{array}{c|ccc} & 1 & 0 & 0 \\ A & 0 & 1 & 0 \\ & 0 & 0 & 1 \end{array}\right] = [A \mid I_3].$$

The procedure ends when this matrix is converted into

$$[I_3 \mid X_1 \quad X_2 \quad X_3].$$

That is, since $A^{-1} = [X_1 \quad X_2 \quad X_3]$, the procedure ends with A^{-1} on the right. This is the reasoning behind the Gauss-Jordan method of calculating inverses.

PRACTICE PROBLEMS 5

1. Use the Gauss-Jordan method to calculate the inverse of the matrix

$$\begin{bmatrix} 1 & 0 & 2 \\ 0 & 1 & -4 \\ 0 & 0 & 2 \end{bmatrix}.$$

2. Solve the system of linear equations

$$\begin{cases} x \phantom{{}+y} + 2z = 4 \\ \phantom{x+{}} y - 4z = 6 \\ \phantom{x+y-{}} 2z = 9. \end{cases}$$

EXERCISES 5

Use the Gauss-Jordan method to compute the inverses of the following matrices.

1. $\begin{bmatrix} 7 & 3 \\ 5 & 2 \end{bmatrix}$

2. $\begin{bmatrix} 5 & -2 \\ 6 & 2 \end{bmatrix}$

3. $\begin{bmatrix} 10 & 12 \\ 3 & -4 \end{bmatrix}$

4. $\begin{bmatrix} 1 & -3 \\ 0 & 1 \end{bmatrix}$

5. $\begin{bmatrix} 2 & -4 \\ -1 & 2 \end{bmatrix}$

6. $\begin{bmatrix} 1 & 3 & 1 \\ -1 & 2 & 0 \\ 2 & 11 & 3 \end{bmatrix}$

7. $\begin{bmatrix} 1 & 2 & -2 \\ 1 & 1 & 1 \\ 0 & 0 & 1 \end{bmatrix}$

8. $\begin{bmatrix} 2 & 2 & 0 \\ 0 & -2 & 0 \\ 3 & 0 & 1 \end{bmatrix}$

9. $\begin{bmatrix} -2 & 5 & 2 \\ 1 & -3 & -1 \\ -1 & 2 & 1 \end{bmatrix}$

10. $\begin{bmatrix} 1 & 0 & 0 \\ 2 & 1 & -2 \\ -1 & 2 & 1 \end{bmatrix}$

11. $\begin{bmatrix} 1 & 6 & 0 & 0 \\ 1 & 5 & 0 & 0 \\ 0 & 0 & 4 & 2 \\ 0 & 0 & 50 & 2 \end{bmatrix}$

12. $\begin{bmatrix} 6 & 0 & 2 & 0 \\ -6 & 1 & 0 & 1 \\ 1 & 0 & 1 & 0 \\ -9 & 0 & -1 & 1 \end{bmatrix}$

In Exercises 13-16, use matrix inversion to solve the system of linear equations.

13. $\begin{cases} x + y + 2z = 3 \\ 3x + 2y + 2z = 4 \\ x + y + 3z = 5 \end{cases}$

14. $\begin{cases} x + 2y + 3z = 4 \\ 3x + 5y + 5z = 3 \\ 2x + 4y + 2z = 4 \end{cases}$

15. $\begin{cases} x \qquad - 2z - 2w = 0 \\ \quad y \qquad - 5w = 1 \\ -4x \qquad + 9z + 9w = 2 \\ \quad 2y + z - 8w = 3 \end{cases}$

16. $\begin{cases} y + 2z = 1 \\ 2x + y + 3z = 2 \\ x + y + 2z = 3 \end{cases}$

SOLUTIONS TO PRACTICE PROBLEMS 5

1. First write the given matrix beside an identity matrix of the same size

$$\left[\begin{array}{ccc|ccc} 1 & 0 & 2 & 1 & 0 & 0 \\ 0 & 1 & -4 & 0 & 1 & 0 \\ 0 & 0 & 2 & 0 & 0 & 1 \end{array}\right].$$

The object is to use elementary row operations to transform the 3×3 matrix on the left into the identity matrix. The first two columns are already in the correct form.

$$\left[\begin{array}{ccc|ccc} 1 & 0 & 2 & 1 & 0 & 0 \\ 0 & 1 & -4 & 0 & 1 & 0 \\ 0 & 0 & 2 & 0 & 0 & 1 \end{array}\right]$$

$$\xrightarrow{\frac{1}{2}[3]} \left[\begin{array}{ccc|ccc} 1 & 0 & 2 & 1 & 0 & 0 \\ 0 & 1 & -4 & 0 & 1 & 0 \\ 0 & 0 & 1 & 0 & 0 & \frac{1}{2} \end{array}\right]$$

$$\xrightarrow{[1] + (-2)[3]} \left[\begin{array}{ccc|ccc} 1 & 0 & 0 & 1 & 0 & -1 \\ 0 & 1 & -4 & 0 & 1 & 0 \\ 0 & 0 & 1 & 0 & 0 & \frac{1}{2} \end{array}\right]$$

$$\xrightarrow{[2] + (4)[3]} \left[\begin{array}{ccc|ccc} 1 & 0 & 0 & 1 & 0 & -1 \\ 0 & 1 & 0 & 0 & 1 & 2 \\ 0 & 0 & 1 & 0 & 0 & \frac{1}{2} \end{array}\right].$$

Thus the inverse of the given matrix is

$$\begin{bmatrix} 1 & 0 & -1 \\ 0 & 1 & 2 \\ 0 & 0 & \frac{1}{2} \end{bmatrix}.$$

2. The matrix form of this system of equations is $AX = B$, where A is the matrix whose inverse was found in Problem 1, and

$$B = \begin{bmatrix} 4 \\ 6 \\ 9 \end{bmatrix}.$$

Therefore, $X = A^{-1}B$, so that

$$\begin{bmatrix} x \\ y \\ z \end{bmatrix} = \begin{bmatrix} 1 & 0 & -1 \\ 0 & 1 & 2 \\ 0 & 0 & \frac{1}{2} \end{bmatrix} \begin{bmatrix} 4 \\ 6 \\ 9 \end{bmatrix} = \begin{bmatrix} -5 \\ 24 \\ \frac{9}{2} \end{bmatrix}.$$

So the solution of the system is $x = -5$, $y = 24$, $z = \frac{9}{2}$.

1.6 Input-Output Analysis

In recent years matrix arithmetic has played an ever-increasing role in economics, especially in that branch of economics called *input-output analysis.* Pioneered by the Harvard economist Vassily Leontieff, input-output analysis is used to analyze an economy in order to determine how much output must be produced by each segment of the economy in order to meet given consumption and export demands. As we shall see, such analysis leads into matrix calculations and in particular to inverses of matrices. Input-output analysis has been of such great significance that Leontieff was awarded the 1973 Nobel prize in economics for his fundamental work in the subject.

Suppose that we divide an economy up into a number of industries—transportation, agriculture, steel, and so on. Each industry produces a certain output using certain raw materials (or input). The input of each industry is made up in part by the outputs of other industries. For example, in order to produce food, agriculture uses as input the output of many industries, such as transportation (tractors and trucks) and oil (gasoline and fertilizers). This interdependence among the industries of the economy is summarized in a matrix—a so-called *input-output matrix.* There is one column for each industry's input requirements. The entries in the column reflect the amount of input required from each of the industries. A typical input-output matrix looks like this:

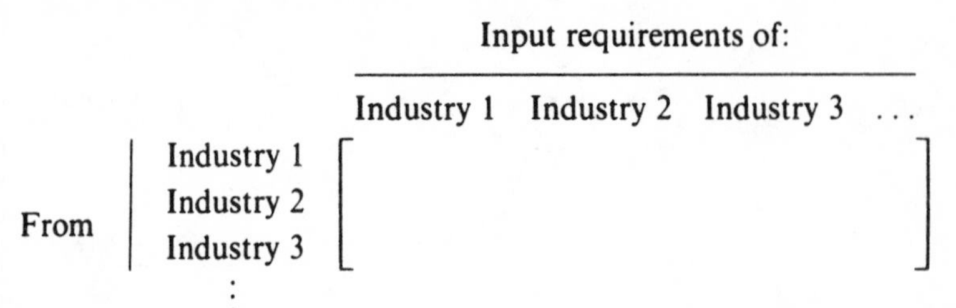

It is most convenient to express the entries of this matrix in monetary terms. That is, each column gives the dollar values of the various inputs needed by an industry in order to produce $1 worth of output.

There are consumers (other than the industries themselves) who want to purchase some of the output of these industries. The quantity of goods that these consumers want (or demand) is called the *final demand* on the economy. The final demand can be represented by a column matrix, with one entry for each industry, indicating the amount of consumable output demanded from the industry:

$$[\text{final demand}] = \begin{bmatrix} \text{amount from industry 1} \\ \text{amount from industry 2} \\ \vdots \end{bmatrix}.$$

We shall consider the situation in which the final-demand matrix is given and it is necessary to determine how much output should be produced by each industry in order to provide the needed inputs of the various industries and also to satisfy the final demand. The proper level of output can be computed using matrix calculations as illustrated in the next example.

EXAMPLE 1 Suppose that an economy is composed of only three industries—coal, steel, and electricity. Each of these industries depends on the others for some of its raw materials. Suppose that to make $1 of coal, it takes no coal, but $.02 of steel and $.01 of electricity; to make $1 of steel, it takes $.15 of coal, $.03 of steel, and $.08 of electricity; and to make $1 of electricity, it takes $.43 of coal, $.20 of steel, and $.05 of electricity. How much should each industry produce to allow for consumption (not used for production) at these levels: $2 billion coal, $1 billion steel, $3 billion electricity?

Solution Put all the data indicating the interdependence of the industries in a matrix. In each industry's column put the amount of input from each of the industries needed to produce $1 of output in that particular industry:

$$\begin{array}{l} \\ \text{Coal} \\ \text{Steel} \\ \text{Electricity} \end{array} \begin{array}{c} \begin{array}{ccc} \text{Coal} & \text{Steel} & \text{Electricity} \end{array} \\ \begin{bmatrix} 0 & .15 & .43 \\ .02 & .03 & .20 \\ .01 & .08 & .05 \end{bmatrix} \end{array} = A.$$

This matrix is the *input-output matrix* corresponding to the economy. Let D denote the final-demand matrix. Then, letting the numbers in D stand for billions of dollars, we have

$$D = \begin{bmatrix} 2 \\ 1 \\ 3 \end{bmatrix}.$$

Suppose that the coal industry produces x billion dollars of output, the steel industry y billion dollars, and the electrical industry z billion dollars. Our problem is to determine the x, y, and z that yield the desired amounts left over from the production process. As an example, consider coal. The amount of coal that can be consumed or exported is just

$$x - [\text{amount of coal used in production}].$$

To determine the amount of coal used in production, refer to the input-output matrix. Production of x billion dollars of coal takes $0 \cdot x$ billion dollars of coal, production of y billion dollars of steel takes $.15y$ billion dollars of coal, and production of z billion dollars of electricty takes $.43z$ billion dollars of coal. Thus,

$$[\text{amount of coal used in production}] = 0 \cdot x + .15y + .43z.$$

This quantity should be recognized as the first entry of a matrix product. Namely, if we let

$$X = \begin{bmatrix} x \\ y \\ z \end{bmatrix},$$

then

$$\begin{bmatrix} \text{coal} \\ \text{steel} \\ \text{electricity} \end{bmatrix}_{\text{used in production}} = \begin{bmatrix} 0 & .15 & .43 \\ .02 & .03 & .20 \\ .01 & .08 & .05 \end{bmatrix} \begin{bmatrix} x \\ y \\ z \end{bmatrix}.$$
$$= AX.$$

But then the amount of each output available for purposes other than production is $X - AX$. That is, we have the matrix equation

$$X - AX = D.$$

To solve this equation for X, proceed as follows. Since $IX = X$, write the equation in the form

$$IX - AX = D$$
$$(I - A)X = D$$

$$\boxed{X = (I - A)^{-1}D.} \qquad (1)$$

So, in other words, X may be found by multiplying D on the left by $(I - A)^{-1}$. Let us now do the arithmetic.

$$I - A = \begin{bmatrix} 1 & 0 & 0 \\ 0 & 1 & 0 \\ 0 & 0 & 1 \end{bmatrix} - \begin{bmatrix} 0 & .15 & .43 \\ .02 & .03 & .20 \\ .01 & .08 & .05 \end{bmatrix} = \begin{bmatrix} 1 & -.15 & -.43 \\ -.02 & .97 & -.20 \\ -.01 & -.08 & .95 \end{bmatrix}.$$

Applying the Gauss-Jordan method, we find

$$(I - A)^{-1} = \begin{bmatrix} 1.01 & .20 & .50 \\ .02 & 1.05 & .23 \\ .01 & .09 & 1.08 \end{bmatrix}.$$

where all figures are carried to two decimal places (Exercise 5). Therefore,

$$X = (I - A)^{-1}D = \begin{bmatrix} 1.01 & .20 & .50 \\ .02 & 1.05 & .23 \\ .01 & .09 & 1.08 \end{bmatrix} \begin{bmatrix} 2 \\ 1 \\ 3 \end{bmatrix} = \begin{bmatrix} 3.72 \\ 1.78 \\ 3.35 \end{bmatrix}.$$

In other words, coal should produce \$3.72 billion worth of output, steel \$1.78 billion, and electricity \$3.35 billion. This output will meet the required final demands from each industry.

The analysis above is useful in studying not only entire economies but also segments of economies and even individual companies.

EXAMPLE 2 A conglomerate has three divisions, which produce computers, semiconductors, and business forms. For each \$1 of output, the computer division needs \$.02 worth of computers, \$.20 worth of semiconductors, and \$.10 worth of business forms. For each \$1 of output, the semiconductor division needs \$.02 worth of computers, \$.01 worth of semiconductors, and \$.02 worth of business forms. For each \$1 of output, the business forms division requires \$.10 worth of computers and \$.01 worth of business forms. The conglomerate estimates the sales demand to be \$300,000,000 for the computer division, \$100,000,000 for the semiconductor division, and \$200,000,000 for the business forms division. At what level should each division produce in order to satisfy this demand?

Solution The conglomerate can be viewed as a miniature economy and its sales as the final demand. The input-output matrix for this "economy" is

	Computers	Semiconductors	Business forms
Computers	.02	.02	.10
Semiconductors	.20	.01	0
Business forms	.10	.02	.01

$= A.$

The final-demand matrix is

$$D = \begin{bmatrix} 3 \\ 1 \\ 2 \end{bmatrix},$$

where the demand is expressed in hundreds of millions of dollars. By equation (1) the matrix X, giving the desired levels of production for the various divisions, is given by

$$X = (I - A)^{-1}D.$$

But

$$I - A = \begin{bmatrix} .98 & -.02 & -.10 \\ -.20 & .99 & 0 \\ -.10 & -.02 & .99 \end{bmatrix},$$

so that (Exercise 6)

$$(I - A)^{-1} = \begin{bmatrix} 1.04 & .02 & .10 \\ .21 & 1.01 & .02 \\ .11 & .02 & 1.02 \end{bmatrix},$$

and

$$(I - A)^{-1}D = \begin{bmatrix} 3.34 \\ 1.68 \\ 2.39 \end{bmatrix}.$$

Therefore,

$$X = \begin{bmatrix} 3.34 \\ 1.68 \\ 2.39 \end{bmatrix}.$$

That is, the computer division should produce \$334,000,000, the semiconductor division \$168,000,000, and the business forms division \$239,000,000.

Input-output analysis is usually applied to the entire economy of a country having hundreds of industries. The resulting matrix equation $(I - A)X = D$ could be solved by the Gaussian elimination method. However, it is best to find the inverse of $I - A$ and solve for X as we have done in the examples of this section. Over a short period, D might change but A is unlikely to change. Therefore, the proper outputs to satisfy the new demand can easily be determined by using the already computed inverse of $I - A$.

The Closed Leontieff Model The foregoing description of an economy is usually called the *Leontieff open model*, since it views exports as an activity that takes place external to the economy. However, it is possible to consider export as but another industry in the economy. Instead of describing exports by a demand column D, we describe it by a column in the input-output matrix. That is, the export column describes how each dollar of exports is divided among the various industries. Since exports are now regarded as another industry, each of the original columns has an additional entry, namely the amount of output from the export industry (that is, imports) used to produce \$1 of goods (of the industry corresponding to the column). If A denotes the expanded input-output matrix and X the production matrix (as before), then AX is the matrix describing the total demand experienced by each of the industries. In order for the economy to function efficiently, the total amount demanded by the various industries should equal the amount produced. That is, the production matrix must satisfy the equation

$$AX = X.$$

By studying the solutions to this equation, it is possible to determine the equilibrium states of the economy—that is, the production matrices X for which the amounts produced exactly equal the amounts needed by the various industries. The model just described is called *Leontieff's closed model.*

We may expand the Leontieff closed model to include the effects of labor and monetary phenomena by considering labor and banking as yet further industries to be incorporated in the input-output matrix.

PRACTICE PROBLEMS 6

1. Let

$$I = \begin{bmatrix} 1 & 0 & 0 \\ 0 & 1 & 0 \\ 0 & 0 & 1 \end{bmatrix}, \quad A = \begin{bmatrix} .1 & 0 & .1 \\ .2 & .1 & .1 \\ .1 & .2 & 0 \end{bmatrix}, \quad X = \begin{bmatrix} x \\ y \\ z \end{bmatrix}, \quad D = \begin{bmatrix} 100 \\ 200 \\ 50 \end{bmatrix}.$$

   Solve the matrix equation

$$(I - A)X = D.$$

2. Let I, A, X be as in Problem 1, but let

$$D = \begin{bmatrix} 300 \\ 100 \\ 100 \end{bmatrix}.$$

   Solve the matrix equation $(I - A)X = D$.

EXERCISES 6

1. Suppose that in the economy of Example 1 the demand for electricity triples and the demand for coal doubles, whereas the demand for steel increases only by 50%. At what levels should the various industries produce in order to satisfy the new demand?
2. Suppose that the conglomerate of Example 2 is faced with an increase of 50% in demand for computers, a doubling in demand for semiconductors, and a decrease of 50% in demand for business forms. At what levels should the various divisions produce in order to satisfy the new demand?
3. Suppose that the conglomerate of Example 2 experiences a doubling in the demand for business forms. At what levels should the computer and semiconductor divisions produce?
4. A multinational corporation does business in the United States, Canada, and England. Its branches in one country purchase goods from the branches in other countries according to the matrix

		Branch in:		
		United States	Canada	England
Purchase from	United States	.02	0	.02
	Canada	.01	.03	.01
	England	.03	0	.01

   where the entries in the matrix represent percentages of total sales by the respective branch. The external sales by each of the offices are $800,000,000 for the U.S. branch,

$300,000,000 for the Canadian branch, and $1,400,000,000 for the English branch. At what level should each of the branches produce in order to satisfy the total demand?

5. Show that to two decimal places

$$\begin{bmatrix} 1 & -.15 & -.43 \\ -.02 & .97 & -.20 \\ -.01 & -.08 & .95 \end{bmatrix}^{-1} = \begin{bmatrix} 1.01 & .20 & .50 \\ .02 & 1.05 & .23 \\ .01 & .09 & 1.08 \end{bmatrix}.$$

(A hand calculator would help a great deal here.)

6. Show that to two decimal places

$$\begin{bmatrix} .98 & -.02 & -.10 \\ -.2 & .99 & 0 \\ -.10 & -.02 & .99 \end{bmatrix}^{-1} = \begin{bmatrix} 1.04 & .02 & .10 \\ .21 & 1.01 & .02 \\ .11 & .02 & 1.02 \end{bmatrix}.$$

7. A corporation has a plastics division and an industrial equipment division. For each $1 worth of output, the plastics division needs $.02 worth of plastics and $.10 worth of equipment. For each $1 worth of output, the industrial equipment division needs $.01 worth of plastics and $.05 worth of equipment. At what level should the divisions produce to meet a demand for $930,000 worth of plastics and $465,000 worth of industrial equipment?

8. Rework Exercise 7 under the condition that the demand for plastics is $1,860,000 and the demand for industrial equipment is $2,790,000.

SOLUTIONS TO PRACTICE PROBLEMS 6

1. The equation $(I - A)X = D$ has the form $CX = D$, where C is the matrix $I - A$. From Section 4 we know that $X = C^{-1}D$. That is,

$$X = (I - A)^{-1}D.$$

Now

$$I - A = \begin{bmatrix} 1 & 0 & 0 \\ 0 & 1 & 0 \\ 0 & 0 & 1 \end{bmatrix} - \begin{bmatrix} .1 & 0 & .1 \\ .2 & .1 & .1 \\ .1 & .2 & 0 \end{bmatrix} = \begin{bmatrix} .9 & 0 & -.1 \\ -.2 & .9 & -.1 \\ -.1 & -.2 & 1 \end{bmatrix}.$$

Using the Gauss-Jordan method to find the inverse of this matrix, we have (to two decimal places)

$$(I - A)^{-1} = \begin{bmatrix} 1.13 & .03 & .12 \\ .27 & 1.14 & .14 \\ .17 & .23 & 1.04 \end{bmatrix}.$$

Therefore, rounding to the nearest integer, we have

$$X = (I - A)^{-1}D = \begin{bmatrix} 1.13 & .03 & .12 \\ .27 & 1.14 & .14 \\ .17 & .23 & 1.04 \end{bmatrix} \begin{bmatrix} 100 \\ 200 \\ 50 \end{bmatrix} = \begin{bmatrix} 125 \\ 262 \\ 115 \end{bmatrix}.$$

2. We have $X = (I - A)^{-1}D$, where $(I - A)^{-1}$ is as computed in Problem 1. So

$$X = \begin{bmatrix} 1.13 & .03 & .12 \\ .27 & 1.14 & .14 \\ .17 & .23 & 1.04 \end{bmatrix} \begin{bmatrix} 300 \\ 100 \\ 100 \end{bmatrix} = \begin{bmatrix} 354 \\ 209 \\ 178 \end{bmatrix}.$$

Chapter 5: CHECKLIST

- □ System of linear equations
- □ Elementary row operations
- □ Diagonal form
- □ Gaussian elimination method
- □ Matrix
- □ Pivoting
- □ Row matrix
- □ Column matrix
- □ Square matrix
- □ The ijth entry, a_{ij}
- □ Addition and subtraction of matrices
- □ Multiplication of matrices
- □ Identity matrix, I_n
- □ Inverse of a matrix, A^{-1}
- □ Formula for inverse of a 2×2 matrix
- □ Solution of matrix equation $AX = B$
- □ Use of inverse matrix to solve system of linear equations
- □ Gauss-Jordan method for calculating inverse of a matrix
- □ Input-output analysis

Chapter 5: SUPPLEMENTARY EXERCISES

Pivot each of the following matrices around the circled element.

1. $\begin{bmatrix} \textcircled{3} & -6 & 1 \\ 2 & 4 & 6 \end{bmatrix}$

2. $\begin{bmatrix} -5 & -3 & 1 \\ 4 & \textcircled{2} & 0 \\ 0 & 6 & 7 \end{bmatrix}$

Use the Gaussian elimination method to find all solutions of the following systems of linear equations.

3. $\begin{cases} \frac{1}{2}x - y = -3 \\ 4x - 5y = -9 \end{cases}$

4. $\begin{cases} 3x + 9z = 42 \\ 2x + y + 6z = 30 \\ -x + 3y - 2z = -20 \end{cases}$

5. $\begin{cases} 3x - 6y + 6z = -5 \\ -2x + 3y - 5z = \frac{7}{3} \\ x + y + 10z = 3 \end{cases}$

6. $\begin{cases} 3x + 6y - 9z = 1 \\ 2x + 4y - 6z = 1 \\ 3x + 4y + 5z = 0 \end{cases}$

7. $\begin{cases} x + 2y - 5z + 3w = 16 \\ -5x - 7y + 13z - 9w = -50 \\ -x + y - 7z + 2w = 9 \\ 3x + 4y - 7z + 6w = 33 \end{cases}$

8. $\begin{cases} 5x - 10y = 5 \\ 3x - 8y = -3 \\ -3x + 7y = 0 \end{cases}$

Perform the indicated matrix operations.

9. $\begin{bmatrix} 2 \\ -1 \\ 0 \end{bmatrix} + \begin{bmatrix} 3 \\ 4 \\ 7 \end{bmatrix}$

10. $\begin{bmatrix} 1 & 3 & -2 \\ 4 & 0 & -1 \end{bmatrix}\begin{bmatrix} 3 & 5 \\ 1 & 0 \\ 0 & -6 \end{bmatrix}$

11. Find the inverse of the appropriate matrix and use it to solve the system of equations

$$\begin{cases} 3x + 2y = 0 \\ 5x + 4y = 2. \end{cases}$$

12. The matrices

$$\begin{bmatrix} 4 & -2 & 3 \\ 8 & -3 & 5 \\ 7 & -2 & 4 \end{bmatrix} \quad \text{and} \quad \begin{bmatrix} -2 & 2 & -1 \\ 3 & -5 & 4 \\ 5 & -6 & 4 \end{bmatrix}$$

are inverses of each other. Use these matrices to solve the following systems of linear equations.

(a) $\begin{cases} -2x + 2y - z = 1 \\ 3x - 5y + 4z = 0 \\ 5x - 6y + 4z = 3 \end{cases}$

(b) $\begin{cases} 4x - 2y + 3z = 0 \\ 8x - 3y + 5z = -1 \\ 7x - 2y + 4z = 2 \end{cases}$

Use the Gauss-Jordan method to calculate the inverses of the following matrices.

13. $\begin{bmatrix} 2 & 6 \\ 1 & 2 \end{bmatrix}$

14. $\begin{bmatrix} 1 & 1 & 1 \\ 3 & 4 & 3 \\ 1 & 1 & 2 \end{bmatrix}$

15. The economy of a small country can be regarded as consisting of two industries, I and II, whose input-output matrix is

$$A = \begin{bmatrix} .4 & .2 \\ .1 & .3 \end{bmatrix}.$$

How many units should be produced by each industry in order to meet a demand for 8 units from industry I and 12 units from industry II?

Linear Programming, A Geometric Approach

Chapter 2

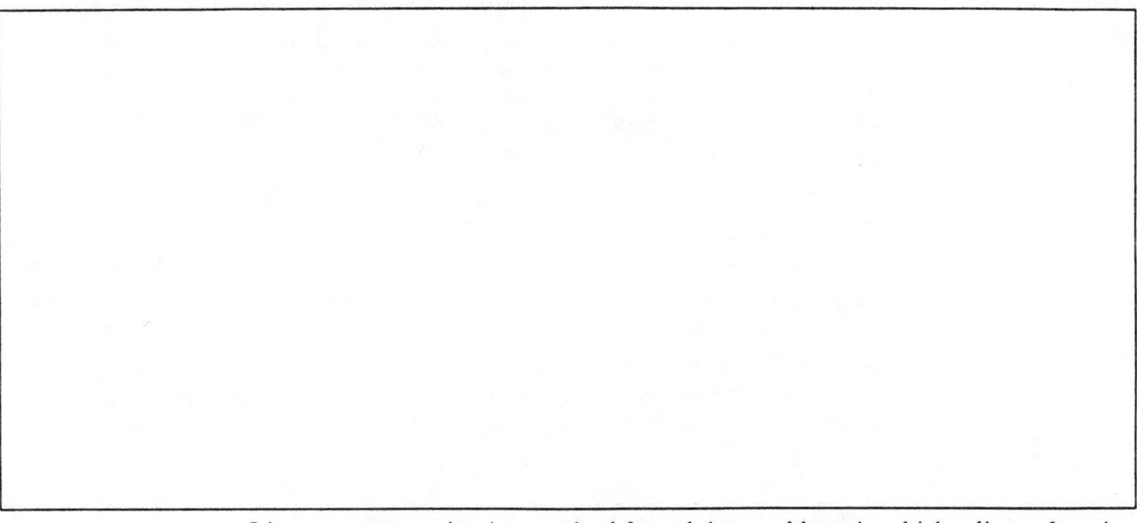

Linear programming is a method for solving problems in which a linear function (representing cost, profit, distance, weight, or the like) is to be maximized or minimized. Such problems are called *optimization problems.* As we shall see, these problems, when translated into mathematical language, involve systems of linear inequalities, systems of linear equations, and eventually (in **Chapter 3**) matrices.

2.1 A Linear Programming Problem

Let us begin with a detailed discussion of a typical problem that can be solved by linear programming.

Furniture Manufacturing Problem

A furniture manufacturer makes two types of furniture—chairs and sofas. For simplicity, divide the production process into three distinct operations—carpentry, finishing, and upholstery. The amount of labor required for each operation varies. Manufacture of a chair requires 6 hours of carpentry, 1 hour of finishing, and 2 hours of upholstery. Manufacture of a sofa requires 3 hours of carpentry, 1 hour of finishing, and 6 hours of upholstery. Owing to limited availability of skilled labor as well as of tools and equipment, the factory has available each day 96 man-hours for carpentry, 18 man-hours for finishing, and 72 man-hours for upholstery. The profit per chair is $80 and the profit per sofa $70. How many chairs and how many sofas should be produced each day in order to maximize the profit?

It is often helpful to tabulate data given in verbal problems. Our first step, then, is to construct a chart.

	Chair	Sofa	Available labor
Carpentry	6 hours	3 hours	96 man-hours
Finishing	1 hour	1 hour	18 man-hours
Upholstery	2 hours	6 hours	72 man-hours
Profit	$80	$70	

The next step is to translate the problem into mathematical language. As you know, this is done by identifying what is unknown and denoting the unknown quantities by letters. Since the problem asks for the optimum number of chairs and sofas to be produced each day, there are two unknowns—the number of chairs produced each day and the number of sofas produced each day. Let x denote the former and y the latter.

In order to achieve a large profit, one need only manufacture a large number of chairs and sofas. But, owing to restricted availability of tools and labor, the factory cannot manufacture an unlimited quantity of furniture. Let us translate the restrictions into mathematical language. Each row of the chart gives one restriction. The first row says that the amount of carpentry required is 6 hours for each chair and 3 hours for each sofa. Also, there are available only 96 man-hours of carpentry per day. We can compute the total number of man-hours of carpentry required per day to produce x chairs and y sofas as follows:

$$\begin{aligned}&[\text{number of man-hours per day of carpentry}]\\&\qquad = (\text{number of hours carpentry per chair})\cdot(\text{number of chairs per day})\\&\qquad\quad + (\text{number of hours carpentry per sofa})\cdot(\text{number of sofas per day})\\&\qquad = 6\cdot x + 3\cdot y.\end{aligned}$$

The requirement that at most 96 man-hours of carpentry be used per day means that x and y must satisfy the inequality

$$6x + 3y \leq 96. \tag{1}$$

The second row of the chart gives a restriction imposed by finishing. Since 1 hour of finishing is required for each chair and sofa, and since at most 18 man-hours of finishing are available per day, the same reasoning as used to derive inequality (1) yields

$$x + y \leq 18. \tag{2}$$

Similarly, the third row of the chart gives the restriction due to upholstery:

$$2x + 6y \leq 72. \tag{3}$$

A further restriction is given by the fact that the numbers of chairs and sofas must be nonnegative:

$$x \geq 0, \qquad y \geq 0. \tag{4}$$

Now that we have written down the restrictions which constrain x and y, let us express the profit (which is to be maximized) in terms of x and y. The profit comes from two sources—chairs and sofas. Therefore,

$$\begin{aligned}[\text{profit}] &= [\text{profit from chairs}] + [\text{profit from sofas}] \\ &= [\text{profit per chair}] \cdot [\text{number of chairs}] \\ &\quad + [\text{profit per sofa}] \cdot [\text{number of sofas}] \\ &= 80x + 70y \end{aligned} \tag{5}$$

Combining (1) to (5), we arrive at the following:

□ **Furniture Manufacturing Problem—Mathematical Formulation** Find numbers x and y for which $80x + 70y$ is as large as possible, and for which all the following inequalities hold simultaneously:

$$\begin{cases} 6x + 3y \leq 96 \\ x + y \leq 18 \\ 2x + 6y \leq 72 \\ x \geq 0, \quad y \geq 0. \end{cases} \tag{6}$$

We may describe this mathematical problem in the following general way. We are required to maximize an expression in a certain number of variables, where the variables are subject to restrictions in the form of one or more inequalities. Problems of this sort are called *mathematical programming problems*. Actually, general mathematical programming problems can be quite involved, and their solutions may require very sophisticated mathematical ideas. However, this is not the case with the furniture manufacturing problem. What makes it a rather simple mathematical programming problem is that both the expression to be maximized and the inequalities are linear. For this reason the furniture manufacturing problem is called a *linear programming problem*. The theory of linear programming is a fairly recent advance in mathematics. It was developed over the last 40 years to deal with the increasingly more complicated problems of our technological society. The 1975 Nobel prize in economics was awarded to Kantorovich and Koopmans for their pioneering work in the field of linear programming.

We will solve the furniture manufacturing problem in Section 2, where we will develop a general technique for handling similar linear programming problems. At this point it is worthwhile to attempt to gain some insights into the problem and possible methods for attacking it.

It seems clear that a factory will operate most efficiently when its labor is fully utilized. Let us therefore take the operations one at a time and determine the conditions on x and y that fully utilize the three kinds of labor. The restriction on carpentry asserts that

$$6x + 3y \leq 96.$$

If x and y were chosen so that $6x + 3y$ is actually *less* than 96, we would leave the carpenters idle some of the time, a waste of labor. Thus, it would seem reasonable to choose x and y to satisfy

$$6x + 3y = 96.$$

Similarly, to utilize all the finishers' time, x and y must satisfy

$$x + y = 18,$$

and to utilize all the upholsterers' time, we must have

$$2x + 6y = 72.$$

Thus, if no labor is to be wasted, then x and y must satisfy the system of equations

$$\begin{cases} 6x + 3y = 96 \\ x + y = 18 \\ 2x + 6y = 72. \end{cases} \tag{7}$$

Let us now graph the three equations of (7), which represent the conditions for full utilization of all forms of labor. (See chart below and Fig. 1.)

Equation	*Standard form*	*x-intercept*	*y-intercept*
$6x + 3y = 96$	$y = -2x + 32$	(16, 0)	(0, 32)
$x + y = 18$	$y = -x + 18$	(18, 0)	(0, 18)
$2x + 6y = 72$	$y = -\frac{1}{3}x + 12$	(36, 0)	(0, 12)

What does Fig. 1 say about the furniture manufacturing problem? Each particular pair of numbers (x, y) is called a *production schedule.* Each of the lines in Fig. 1 gives the production schedules which fully utilize one of the types of labor. Notice that the three lines do not have a common intersection point. This means that there is *no* production schedule which *simultaneously* makes full use of all three types of labor. In any production schedule at least some of the man-hours

FIGURE 1

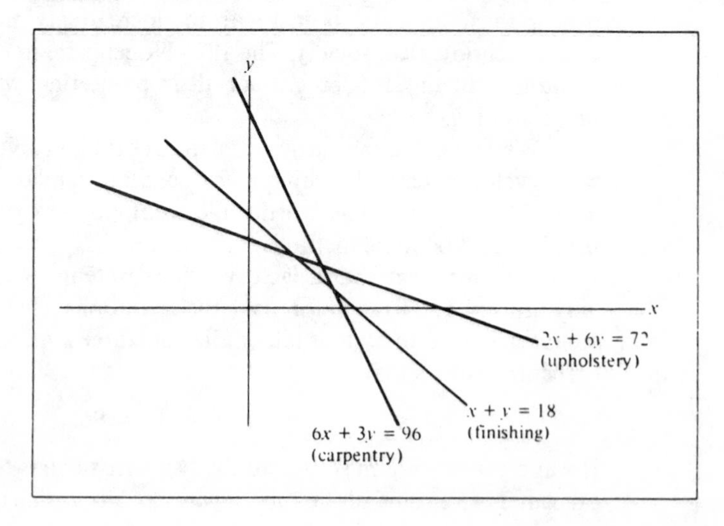

of labor must be wasted. This is not a solution to the furniture manufacturing problem, but is a valuable insight. It says that in the inequalities of (6) not all of the corresponding equations can hold. This suggests that we take a closer look at the system of inequalities.

The standard forms of the inequalities (6) are

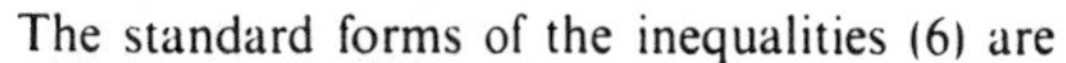

$$\begin{cases} y \le -2x + 32 \\ y \le -x + 18 \\ y \le -\frac{1}{3}x + 12 \\ x \ge 0, \quad y \ge 0. \end{cases}$$

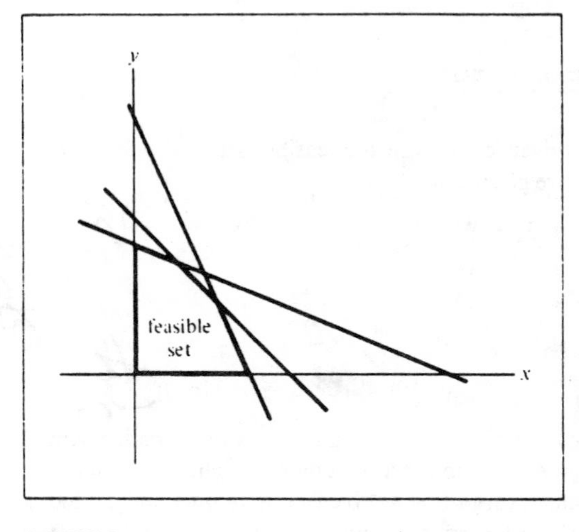

FIGURE 2

By using the techniques of Section 2.2,* we arrive at a feasible set for the system of inequalities above, as shown in Fig. 2.

The feasible set for the furniture manufacturing problem is a bounded, five-sided region. The points on and inside the boundary of this feasible set give the production schedules which satisfy all the restrictions. In the next section we will show how to pick out the particular point of the feasible set that corresponds to a maximum profit.

PRACTICE PROBLEMS 1

1. Determine whether the following points are in the feasible set of the furniture manufacturing problem: (a) (10, 9); (b) (14, 4).

2. A physical fitness enthusiast decides to devote her exercise time to a combination of jogging and cycling. She wants to earn aerobic points (a measure of the benefit of the exercise to strengthening the heart and lungs) and also to achieve relaxation and enjoyment. She jogs at 6 miles per hour and cycles at 18 miles per hour. An hour of jogging earns 12 aerobic points and an hour of cycling earns 9 aerobic points. Each week she would like to earn at least 36 aerobic points, cover at least 54 miles, and cycle at least as much as she jogs.

   (a) Fill in the chart below.

	One hour of jogging	*One hour of cycling*	*Requirement*
Miles covered			
Aerobic points			
Time required			

   (b) Let x be the number of hours of jogging and y the number of hours of cycling each week. Referring to the chart, give the inequalities that x and y must satisfy due to miles covered and aerobic points.

* The reader is referred to ***Mathematics for the Management, Life, and Social Sciences*** **by Goldstein, Lay, and Schneider**

(c) Give the inequalities that x and y must satisfy due to her preference for cycling and also due to the fact that x and y cannot be negative.

(d) Express the time required as a linear function of x and y.

(e) Graph the feasible set for the system of linear inequalities.

EXERCISES 1

In Exercises 1–4, determine whether the given point is in the feasible set of the furniture manufacturing problem. (The inequalities are given below.)

$$\begin{cases} 6x + 3y \le 96 \\ x + y \le 18 \\ 2x + 6y \le 72 \\ x \ge 0, \quad y \ge 0. \end{cases}$$

1. (8, 7) **2.** (14, 3) **3.** (9, 10) **4.** (16, 0)

5. (*Shipping Problem*) A truck traveling from New York to Baltimore is to be loaded with two types of cargo. Each crate of cargo A is 4 cubic feet in volume, weights 100 pounds, and earns $13 for the driver. Each crate of cargo B is 3 cubic feet in volume, weighs 200 pounds, and earns $9 for the driver. The truck can carry no more than 300 cubic feet of crates and no more than 10,000 pounds. Also, the number of crates of cargo B must be less than or equal to twice the number of crates of cargo A.

(a) Fill in the chart below.

	A	*B*	*Truck capacity*
Volume			
Weight			
Earnings			

(b) Let x be the number of crates of cargo A and y the number of crates of cargo B. Referring to the chart, give the two inequalities that x and y must satisfy because of the truck's capacity for volume and weight.

(c) Give the inequalities that x and y must satisfy because of the last sentence of the problem and also because x and y cannot be negative.

(d) Express the earnings from carrying x crates of cargo A and y crates of cargo B.

(e) Graph the feasible set for the shipping problem.

6. (*Mining Problem*) A coal company owns mines in two different locations. Each day mine 1 produces 4 tons of anthracite (hard coal), 4 tons of ordinary coal, and 7 tons of bituminous (soft) coal. Each day mine 2 produces 10 tons of anthracite, 5 tons of ordinary coal, and 5 tons of bituminous coal. It costs the company $150 per day to operate mine 1 and $200 per day to operate mine 2. An order is received for 80 tons of anthracite, 60 tons of ordinary coal, and 75 tons of bituminous coal.

(a) Fill in the chart below.

	Mine 1	*Mine 2*	*Ordered*
Anthracite			
Ordinary			
Bituminous			
Daily cost			

(b) Let x be the number of days mine 1 should be operated and y the number of days mine 2 should be operated. Refer to the chart and give three inequalities that x and y must satisfy to fill the order.

(c) Give other requirements that x and y must satisfy.

(d) Find the cost of operating mine 1 for x days and mine 2 for y days.

(e) Graph the feasible set for the mining problem.

SOLUTIONS TO PRACTICE PROBLEMS 1

1. A point is in the feasible set of a system of inequalities if it satisfies every inequality. Either the original form or the standard form of the inequalities may be used. The original form of the inequalities of the furniture manufacturing problem is

$$\begin{cases} 6x + 3y \leq 96 \\ x + y \leq 18 \\ 2x + 6y \leq 72 \\ x \geq 0, \quad y \geq 0. \end{cases}$$

(a) (10, 9) $\begin{cases} 6(10) + 3(9) \leq 96 \\ 10 + 9 \leq 18 \\ 2(10) + 6(9) \leq 72 \\ 10 \geq 0, \quad 9 \geq 0; \end{cases}$ $\begin{cases} 87 \leq 96 & \text{true} \\ 19 \leq 18 & \text{false} \\ 74 \leq 72 & \text{false} \\ 10 \geq 0, \quad 9 \geq 0. & \text{true} \end{cases}$

(b) (14, 4) $\begin{cases} 6(14) + 3(4) \leq 96 \\ 14 + 4 \leq 18 \\ 2(14) + 6(4) \leq 72 \\ 14 \geq 0, \quad 4 \geq 0; \end{cases}$ $\begin{cases} 96 \leq 96 & \text{true} \\ 18 \leq 18 & \text{true} \\ 52 \leq 72 & \text{true} \\ 14 \geq 0, \quad 4 \geq 0. & \text{true} \end{cases}$

Therefore, (14, 4) is in the feasible set and (10, 9) is not.

2. (a)

	One hour of jogging	*One hour of cycling*	*Requirement*
Miles covered	6	18	54
Aerobic points	12	9	36
Time required	1	1	

(b) Miles covered: $6x + 18y \geq 54$.
Aerobic points: $12x + 9y \geq 36$.

(c) $y \geq x, x \geq 0$. It is not necessary to list $y \geq 0$ since this is automatically assured if the other two inequalities hold.

(d) $x + y$. (An objective of the exercise program might be to minimize $x + y$.)

(e)

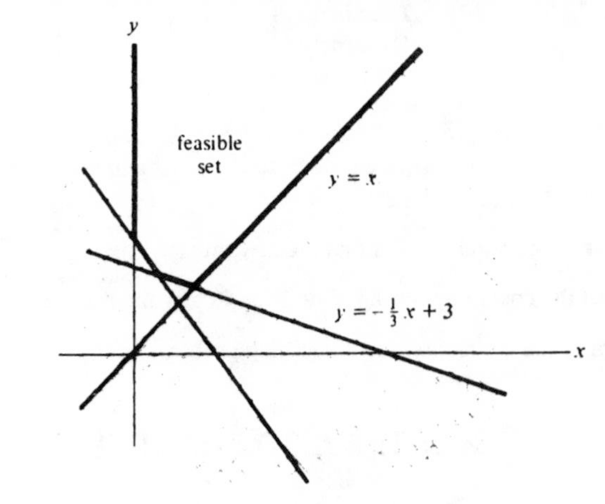

2.2 Linear Programming, I

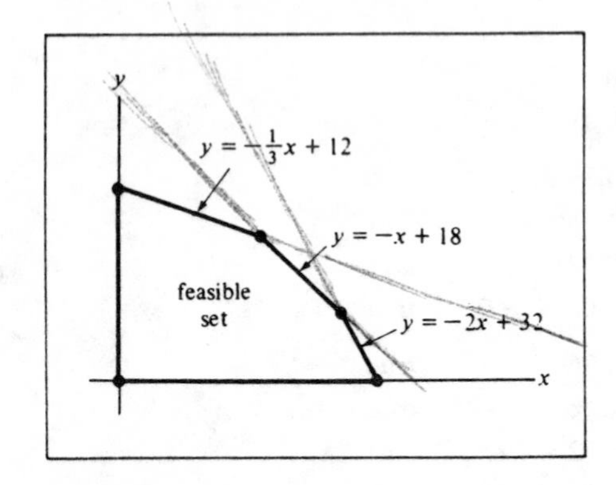

FIGURE 1

We have shown that the feasible set for the furniture manufacturing problem—that is, the set of points corresponding to production schedules satisfying all five restriction inequalities—consists of the points in and on the interior and boundary of the five-sided region drawn in Fig. 1. For reference, we have labeled each line segment with the equation of the line to which it belongs. The line segments intersect in five points, each of which is a corner of the feasible set. Such a corner is called a *vertex*. Somehow, we must pick out of the feasible set an *optimum point*—that is, a point corresponding to a production schedule which yields a maximum profit. To assist us in this task, we have the following result:*

□ **Fundamental Theorem of Linear Programming** The maximum (or minimum) value of the objective function is achieved at one of the vertices of the feasible set.

This result does not completely solve the furniture manufacturing problem for us, but it comes close. It tells us that an optimum production schedule (a, b)

* For a proof, see Section 6.3.

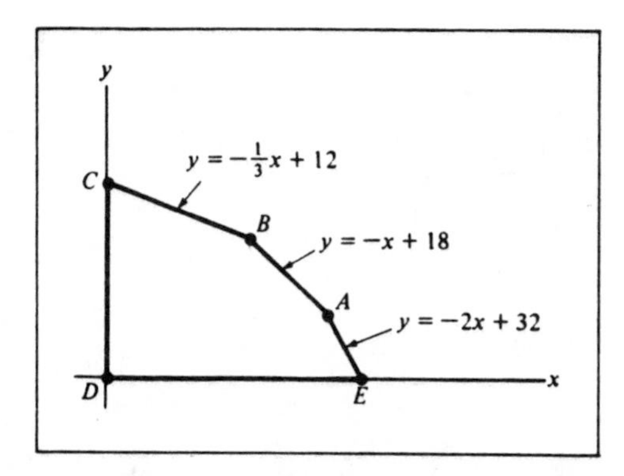

FIGURE 2

corresponds to one of the five points labeled A–E in Fig. 2. So to complete the solution of the furniture manufacturing problem, it suffices to find the coordinates of the five points, evaluate the profit at each, and then choose the point corresponding to the maximum profit.

Solution of the Furniture Manufacturing Problem

Let us begin by determining the coordinates of the points A–E in Fig. 2. Remembering that the x-axis has the equation $y = 0$ and the y-axis the equation $x = 0$, we see from Fig. 2 that the coordinates of A–E can be found as intersections of the following lines:

$$A: \begin{cases} y = -x + 18 \\ y = -2x + 32 \end{cases} \qquad B: \begin{cases} y = -x + 18 \\ y = -\frac{1}{3}x + 12 \end{cases}$$

$$C: \begin{cases} y = -\frac{1}{3}x + 12 \\ x = 0 \end{cases} \qquad D: \begin{cases} y = 0 \\ x = 0 \end{cases}$$

$$E: \begin{cases} y = 0 \\ y = -2x + 32. \end{cases}$$

The point D is clearly $(0, 0)$, and C is clearly the point $(0, 12)$. We obtain A from

$$-x + 18 = -2x + 32$$
$$x = 14$$
$$y = -14 + 18 = 4.$$

Hence $A = (14, 4)$. Similarly, we obtain B from

$$-x + 18 = -\tfrac{1}{3}x + 12$$
$$-\tfrac{2}{3}x = -6$$
$$x = 9$$
$$y = -9 + 18 = 9,$$

so $B = (9, 9)$. Finally, E is obtained from

$$0 = -2x + 32$$
$$x = 16$$
$$y = 0,$$

and thus $E = (16, 0)$. We have listed the vertices in Table 1. In the second column we have evaluated the profit, which is given by $80x + 70y$, at each of the vertices.

TABLE 1

Vertex	*Profit* $= 80x + 70y$
(14, 4)	$80(14) + 70(4) = 1400$
(9, 9)	$80(9) + 70(9) = 1350$
(0, 12)	$80(0) + 70(12) = 840$
(0, 0)	$80(0) + 70(0) = 0$
(16, 0)	$80(16) + 70(0) = 1280$

Note that the largest profit occurs at the vertex (14, 4), so the solution of the linear programming problem is $x = 14$, $y = 4$. In other words, the factory should produce 14 chairs and 4 sofas each day in order to achieve maximum profit, and the maximum profit is $1400 per day.

The furniture manufacturing problem is one particular example of a linear programming problem. Generally, such problems involve finding the values of x and y which maximize (or minimize) a particular linear expression in x and y and where x and y are chosen so as to satisfy one or more restrictions in the form of linear inequalities. The expression that is to be maximized (or minimized) is called the *objective function.* On the basis of our experience with the furniture manufacturing problem, we can summarize the steps to be followed in approaching *any* linear programming problem.

Step 1 Translate the problem into mathematical language.

- **A.** Organize the data.
- **B.** Identify the unknown quantities and define corresponding variables.
- **C.** Translate the restrictions into linear inequalities.
- **D.** Form the objective function.

Step 2 Graph the feasible set.

- **A.** Put the inequalities in standard form.
- **B.** Graph the straight line corresponding to each inequality.
- **C.** Determine the side of the line belonging to the graph of each inequality. Cross out the other side. The remaining region is the feasible set.

Step 3 Determine the vertices of the feasible set.

Step 4 Evaluate the objective function at each vertex. Determine the optimum point.

Linear programming can be applied to many problems. The Army Corps of Engineers has used linear programming to plan the location of a series of dams so as to maximize the resulting hydroelectric power production. The restrictions were to provide adequate flood control and irrigation. Public transit companies have used linear programming to plan routes and schedule buses in order to maximize services. The restrictions in this case arose from the limitations on manpower, equipment, and funding. The petroleum industry uses linear programming in the refining and blending of gasoline. Profit is maximized subject to restrictions on availability of raw materials, refining capacity, and product specifications. Some large advertising firms have used linear programming in media selection. The problem consists of determining how much to spend in each medium in order to maximize the number of consumers reached. The restrictions come from limitations on the budget and the relative costs of different media. Linear programming has been also used by psychologists to design an optimum battery of tests. The problem is to maximize the correlation between test scores and the characteristic that is to be predicted. The restrictions are imposed by the length and cost of the testing.

Linear programming is also used by dieticians in planning meals for large numbers of people. The object is to minimize the cost of the diet, and the restrictions reflect the minimum daily requirements of the various nutrients considered in the diet. The next example is representative of this type of problem. Whereas in actual practice many nutritional factors are considered, we shall simplify the problem by considering only three: protein, calories, and riboflavin.

EXAMPLE 1 (*Nutrition Problem*) Suppose that in a developing nation the government wants to encourage everyone to make rice and soybeans part of his staple diet. The object is to design a lowest-cost diet which provides certain minimum levels of protein, calories, and vitamin B_2 (riboflavin). Suppose that one cup of uncooked rice costs 21 cents and contains 15 grams of protein, 810 calories, and $\frac{1}{9}$ milligram of riboflavin. On the other hand, one cup of uncooked soybeans costs 14 cents and contains 22.5 grams of protein, 270 calories, and $\frac{1}{3}$ millgram of riboflavin. Suppose that the minimum daily requirements are 90 grams of protein, 1620 calories, and 1 milligram of riboflavin. Design the lowest-cost diet meeting these specifications.

Solution We solve the problem by following steps 1–4. The first step is to translate the problem into mathematical language, and the first part of this step is to organize the data, preferably into a chart (Table 2).

Now that we have organized the data, we ask for the unknowns. We wish to know how many cups each of rice and soybeans should comprise the diet, so we identify appropriate variables:

$$x = \text{number of cups of rice per day}$$

$$y = \text{number of cups of soybeans per day.}$$

Next, we obtain the restrictions on the variables. There is one restriction corresponding to each nutrient. That is, there is one restriction for each row of the

TABLE 2

	Rice	Soybeans	Required level per day
Protein (grams/cup)	15	22.5	90
Calories (per cup)	810	270	1620
Riboflavin (milligrams/cup)	$\frac{1}{9}$	$\frac{1}{3}$	1
Cost (cents/cup)	21	14	

chart. If x cups of rice and y cups of soybeans are consumed, then the amount of protein is $15x + 22.5y$ grams. Thus, from the first row of the chart, $15x + 22.5y \geq 90$, a restriction expressing the fact that there must be at least 90 grams of protein per day. Similarly, the restrictions for calories and riboflavin lead to the inequalities $810x + 270y \geq 1620$ and $\frac{1}{9}x + \frac{1}{3}y \geq 1$, respectively. As in the furniture manufacturing problem, x and y cannot be negative, so there are two further restrictions: $x \geq 0$, $y \geq 0$. In all there are five restrictions:

$$\begin{cases} 15x + 22.5y \geq 90 \\ 810x + 270y \geq 1620 \\ \frac{1}{9}x + \frac{1}{3}y \geq 1 \\ x \geq 0, \quad y \geq 0. \end{cases} \tag{1}$$

Now that we have the restrictions, we form the objective function, which tells what we are out to maximize or minimize. Since we wish to minimize cost, we express cost in terms of x and y. Now x cups of rice costs $21x$ cents and y cups of soybeans costs $14y$ cents, so the objective function is given by

$$[\text{cost}] = 21x + 14y. \tag{2}$$

And the problem can finally be stated in mathematical form: Minimize the objective function (2) subject to the restrictions (1). This completes the first step of the solution process.

The second step requires that we graph each of the inequalities (1). In Table 3 we have summarized all the steps necessary to obtain the information from which to draw the graphs. We have sketched the graphs in Fig. 3. From Fig. 3(b) we see

TABLE 3

Inequality	Standard form	Line	Intercepts x	Intercepts y	Graph
$15x + 22.5y \geq 90$	$y \geq -\frac{2}{3}x + 4$	$y = -\frac{2}{3}x + 4$	(6, 0)	(0, 4)	above
$810x + 270y \geq 1620$	$y \geq -3x + 6$	$y = -3x + 6$	(2, 0)	(0, 6)	above
$\frac{1}{9}x + \frac{1}{3}y \geq 1$	$y \geq -\frac{1}{3}x + 3$	$y = -\frac{1}{3}x + 3$	(9, 0)	(0, 3)	above
$x \geq 0$	$x \geq 0$	$x = 0$	(0, 0)	—	right
$y \geq 0$	$y \geq 0$	$y = 0$	—	(0, 0)	above

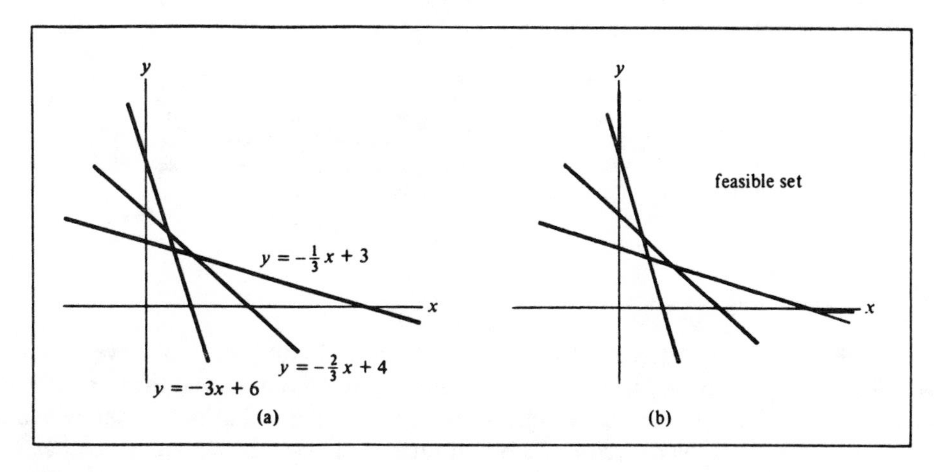

FIGURE 3

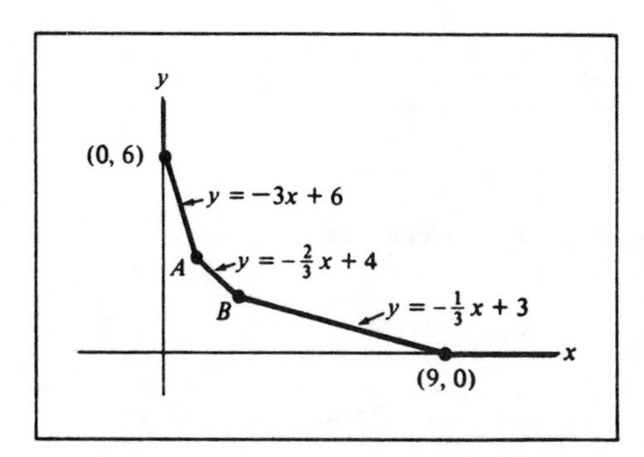

FIGURE 4

that the feasible set is an unbounded, five-sided region. There are four vertices, two of which are known from Table 3, since they are intercepts of boundary lines. Label the remaining two vertices A and B (Fig. 4).

The third step of the solution process consists of determining the coordinates of A and B. From Fig. 4, these coordinates can be found by solving the following systems of equations:

$$A: \begin{cases} y = -3x + 6 \\ y = -\frac{2}{3}x + 4 \end{cases} \qquad B: \begin{cases} y = -\frac{2}{3}x + .4 \\ y = -\frac{1}{3}x + 3. \end{cases}$$

To solve the first system, equate the two expressions for y:

$$-\tfrac{2}{3}x + 4 = -3x + 6$$
$$3x - \tfrac{2}{3}x = 6 - 4$$
$$\tfrac{7}{3}x = 2$$
$$x = \tfrac{6}{7}$$
$$y = -3x + 6 = -3(\tfrac{6}{7}) + 6 = \tfrac{24}{7}$$
$$A = (\tfrac{6}{7}, \tfrac{24}{7}).$$

Similarly, we find B:

$$-\tfrac{2}{3}x + 4 = -\tfrac{1}{3}x + 3$$
$$-\tfrac{1}{3}x = -1$$
$$x = 3$$
$$y = -\tfrac{2}{3}(3) + 4 = 2$$
$$B = (3, 2).$$

TABLE 4

Vertex	*Cost* $= 21x + 14y$
(0, 6)	$21 \cdot 0 + 14 \cdot 6 = 84$
$(\frac{6}{7}, \frac{24}{7})$	$21 \cdot \frac{6}{7} + 14 \cdot \frac{24}{7} = 66$
(3, 2)	$21 \cdot 3 + 14 \cdot 2 = 91$
(9, 0)	$21 \cdot 9 + 14 \cdot 0 = 189$

The fourth step consists of evaluating the objective function, in this case $21x + 14y$, at each vertex. From Table 4, we see that the minimum cost is achieved at the vertex $(\frac{6}{7}, \frac{24}{7})$. So the optimum diet—that is, the one which gives nutrients at the desired levels but at minimum cost—is the one which has $\frac{6}{7}$ cup of rice per day and $\frac{24}{7}$ cups of soybeans per day.

PRACTICE PROBLEMS 2

1. The feasible set for the nutrition problem is shown in the accompanying sketch. The cost is $21x + 14y$. *Without* using the fundamental theorem of linear programming, explain why the cost could not possibly be minimized at the point (4, 4).

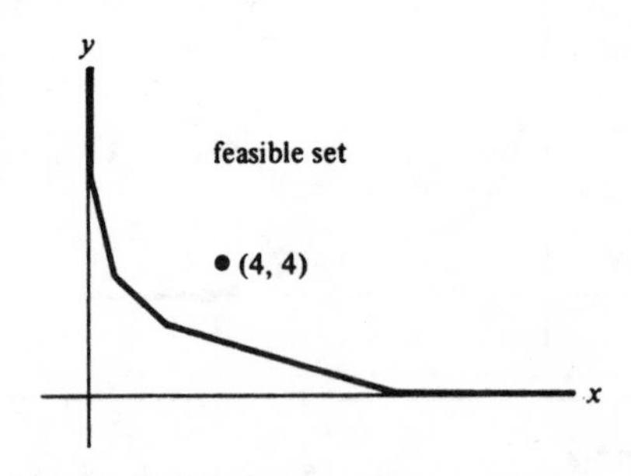

2. Rework the nutrition problem assuming that the cost of rice is changed to 7 cents per cup.

EXERCISES 2

For each of the feasible sets in Exercises 1–4, determine x and y so that the objective function $4x + 3y$ is maximized.

1.

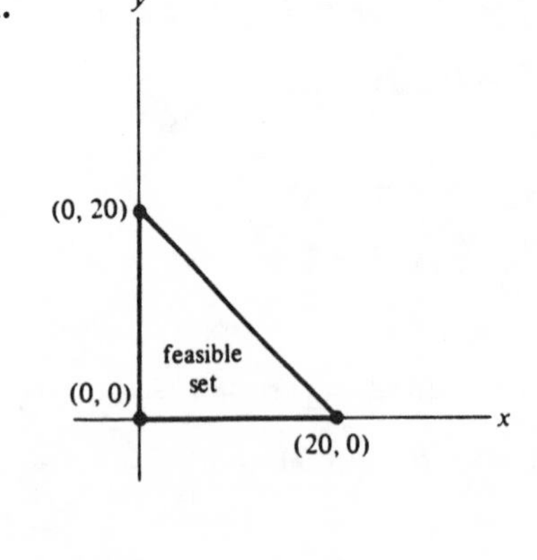

2.

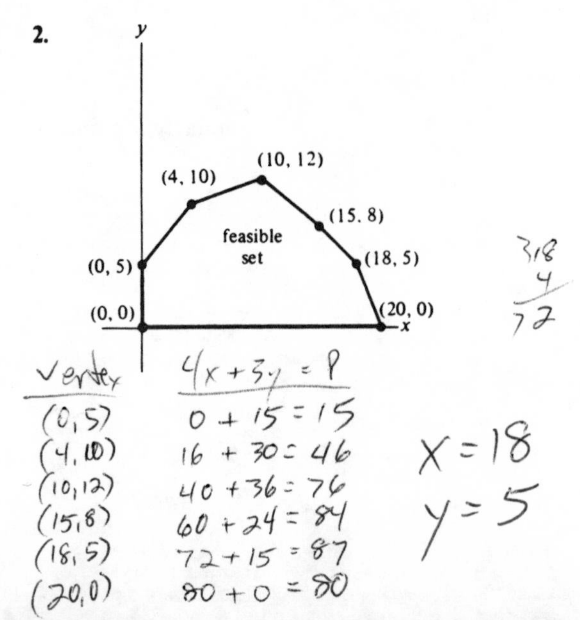

Vertex	$4x+3y=P$
(0, 4)	0 + 12 = 12
(4, 2)	16 + 6 = 22
(6, 0)	24 + 0 = 24

x = 6
y = 0

3. **4.**

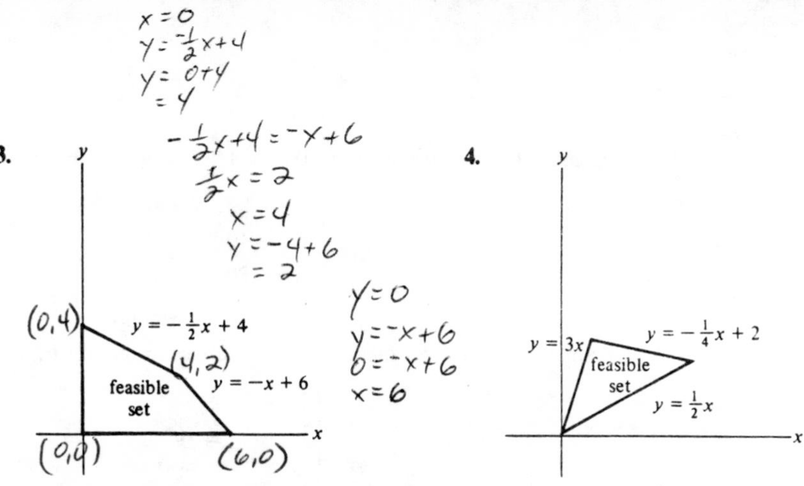

5. (*Shipping Problem*) Refer to Exercises 1, Problem 5. How many crates of each cargo should be shipped in order to satisfy the shipping requirements and yield the greatest earnings? (See the graph of the feasible set below.)

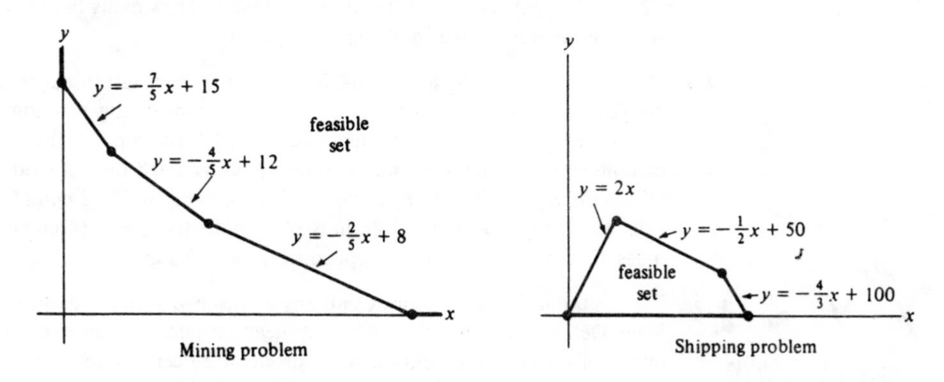

6. (*Mining Problem*) Refer to Exercises 1, Problem 6. Find the number of days that each mine should be operated in order to fill the order at the least cost. (See the graph of the feasible set above.)

In Exercises 7 and 8, rework the furniture manufacturing problem, where everything is the same except that the profit per chair is changed to the given value. (See Table 1 for vertices.)

7. \$150 8. \$60

9. Minimize the objective function $7x + 4y$ subject to the restrictions

$$\begin{cases} y \geq -2x + 11 \\ y \leq -x + 10 \\ y \leq -\frac{1}{3}x + 6 \\ y \geq -\frac{1}{4}x + 4. \end{cases}$$

10. Maximize the objective function $x + 2y$ subject to the restrictions

$$\begin{cases} y \leq -x + 100 \\ y \geq \frac{1}{3}x + 20 \\ y \leq x. \end{cases}$$

11. Maximize the objective function $100x + 150y$ subject to the constraints

$$\begin{cases} x + 3y \le 120 \\ 35x + 10y \le 780 \\ x \le 20 \\ x \ge 0, \quad y \ge 0. \end{cases}$$

12. Minimize the objective function $\frac{1}{2}x + \frac{3}{4}y$ subject to the constraints

$$\begin{cases} 2x + 2y \ge 8 \\ 3x + 5y \ge 16 \\ x \ge 0, \quad y \ge 0. \end{cases}$$

13. A contractor builds two types of homes. The first type requires one lot, \$12,000 capital, and 150 man-days of labor to build and is sold for a profit of \$2400. The second type of home requires one lot, \$32,000 capital, and 200 man-days of labor to build and is sold for a profit of \$3400. The contractor owns 150 lots and has available for the job \$2,880,000 capital and 24,000 man-days of labor. How many homes of each type should she build in order to realize the greatest profit?

14. A nutritionist, working for NASA, must meet certain nutritional requirements and yet keep the weight of the food at a minimum. He is considering a combination of two foods which are packaged in tubes. Each tube of food A contains 4 units of protein, 2 units of carbohydrate, 2 units of fat, and weighs 3 pounds. Each tube of food B contains 3 units of protein, 6 units of carbohydrate, 1 unit of fat, and weighs 2 pounds. The requirement calls for 42 units of protein, 30 units of carbohydrate, and 18 units of fat. How many tubes of each food should be supplied to the astronauts?

15. The Beautiful Day Fruit Juice Company makes two varieties of fruit drink. Each can of Fruit Delight contains 10 ounces of pineapple juice, 3 ounces of orange juice, and 1 ounce of apricot juice, and makes a profit of 20 cents. Each can of Heavenly Punch contains 10 ounces of pineapple juice, 2 ounces of orange juice, and 2 ounces of apricot juice, and makes a profit of 30 cents. Each week, the company has available 9000 ounces of pineapple juice, 2400 ounces of orange juice, and 1400 ounces of apricot juice. How many cans of Fruit Delight and of Heavenly Punch should be produced each week in order to maximize profits?

16. The Bluejay Lacrosse Stick Company makes two kinds of lacrosse sticks. Type A sticks require 2 man-hours for cutting, 1 man-hour for stringing, and 2 man-hours for finishing, and are sold for a profit of \$8. Type B sticks require 1 man-hour for cutting, 3 man-hours for stringing, and 2 man-hours for finishing, and are sold for a profit of \$10. Each day the company has available 120 man-hours for cutting, 150 man-hours for stringing, and 140 man-hours for finishing. How many lacrosse sticks of each kind should be manufactured each day in order to maximize profits?

SOLUTIONS TO PRACTICE PROBLEMS 2

1. The point P has a smaller value of x and a smaller value of y than $(4, 4)$ and is still in the feasible set. It therefore corresponds to a lower cost than $(4, 4)$ and still meets the requirements. We conclude that no interior point of the feasible set could possibly be an

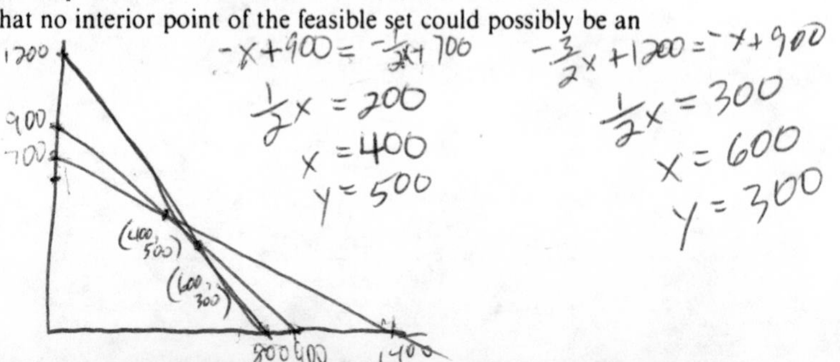

optimum point. This geometric argument indicates that an optimum point might be one that juts out far—that is, a vertex.

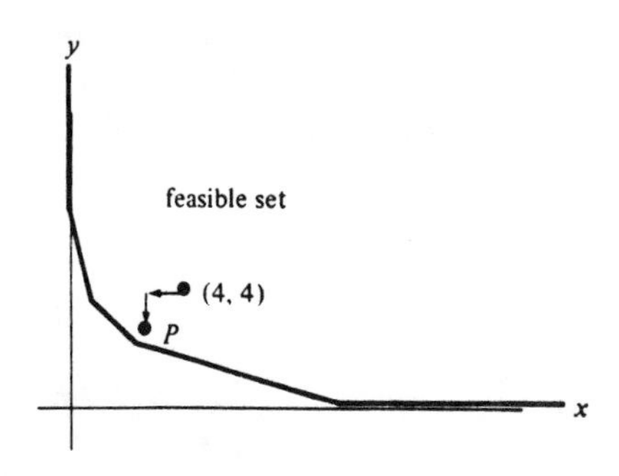

2. The system of linear inequalities, feasible set, and vertices will all be the same as before. Only the objective function changes. The new objective function is $7x + 14y$. The minimum cost occurs when using 3 cups of rice and 2 cups of soybeans.

Vertex	*Cost* $= 7x + 14y$
(0, 6)	84
($\frac{6}{7}$, $\frac{24}{7}$)	54
(3, 2)	49
(9, 0)	63

2.3 Linear Programming, II

In this section we apply the technique of linear programming to the design of a portfolio for a retirement fund and to the transportation of goods from warehouses to retail outlets. The significant new feature of each of these problems is that, on the surface, they appear to involve more than two variables. However, they can be translated into mathematical language so that only two variables are required.

EXAMPLE 1 (*Investment Analysis*) A pension fund has \$30 million to invest. The money is to be divided among Treasury notes, bonds, and stocks. The rules for administration of the fund require that at least \$3 million be invested in each type of investment, at least half the money be invested in Treasury notes and bonds, and the amount invested in bonds not exceed twice the amount invested in Treasury notes. The annual yields for the various investments are 7% for Treasury notes, 8% for bonds, and 9% for stocks. How should the money be allocated among the various investments to produce the largest return?

Solution First, let us agree that all numbers stand for millions. That is, we write 30 to stand for 30 million. This will save us from writing too many zeros. In examining the problem, we find that very little organization needs to be done. The rules for administration of the fund are written in a form from which inequalities can be read right off. Let us just summarize the remaining data in the first row of a chart (Table 1).

TABLE 1

	Treasury notes	*Bonds*	*Stocks*
Yield	.07	.08	.09
Variables	x	y	$30 - (x + y)$

There appear to be three variables—the amounts to be invested in each of the three categories. However, since the three investments must total 30, we need only two variables. Let $x =$ the amount to be invested in Treasury notes and $y =$ the amount to be invested in bonds. Then the amount invested in stocks is $30 - (x + y)$. We have displayed the variables in Table 1.

Now for the restrictions. Since at least 3 (million dollars) must be invested in each category, we have the three inequalities

$$x \geq 3$$

$$y \geq 3$$

$$30 - (x + y) \geq 3.$$

Moreover, since at least half the money, or 15, must be invested in Treasury notes and bonds, we must have

$$x + y \geq 15.$$

Finally, since the amount invested in bonds must not exceed twice the amount invested in Treasury notes, we must have

$$y \leq 2x.$$

(In this example we do not need to state that $x \geq 0$, $y \geq 0$, since we have already required that they be greater than or equal to 3.) Thus there are five restriction inequalities:

$$\begin{cases} x \geq 3, \quad y \geq 3 \\ 30 - (x + y) \geq 3 \\ x + y \geq 15 \\ y \leq 2x. \end{cases} \qquad (1)$$

Next, we form the objective function, which in this case equals the total return on the investment. Since x dollars is invested at 7%, y dollars at 8%, and $30 - (x + y)$ dollars at 9%, the total return is

$$\begin{aligned} [\text{return}] &= .07x + .08y + .09[30 - (x + y)] \\ &= .07x + .08y + 2.7 - .09x - .09y \\ &= 2.7 - .02x - .01y. \end{aligned} \qquad (2)$$

So the mathematical statement of the problem is: Maximize the objective function (2) subject to the restrictions (1).

TABLE 2

Inequality	*Standard form*	*Equation*	*Intercepts* x	y	*Graph*
$x \geq 3$	$x \geq 3$	$x = 3$	(3, 0)	—	Right of line
$y \geq 3$	$y \geq 3$	$y = 3$	—	(0, 3)	Above line
$30 - (x + y) \geq 3$	$y \leq -x + 27$	$y = -x + 27$	(27, 0)	(0, 27)	Below line
$x + y \geq 15$	$y \geq -x + 15$	$y = -x + 15$	(15, 0)	(0, 15)	Above line
$y \leq 2x$	$y \leq 2x$	$y = 2x$	(0, 0)	(0, 0)	Below line

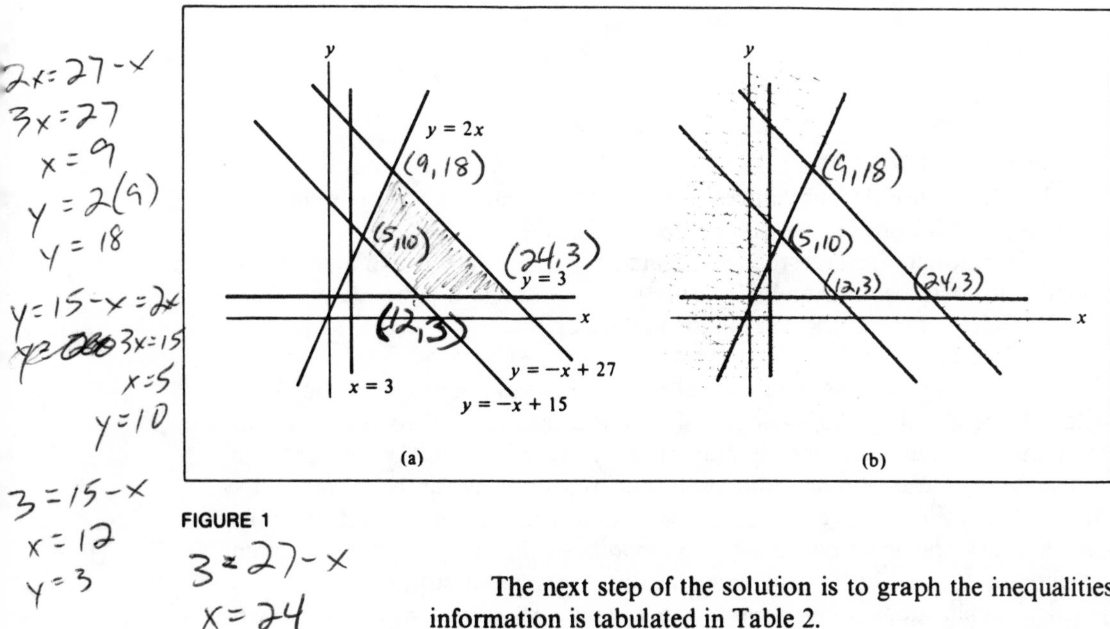

FIGURE 1

The next step of the solution is to graph the inequalities (1). The necessary information is tabulated in Table 2.

One point about the chart is worth noting: It contains enough data to graph each of the lines, with the exception of $y = 2x$. The reason is that the x- and y-intercepts of this line are the same, (0, 0). So to graph $y = 2x$, we must find an additional point on the line. For example, if we set $x = 2$, then $y = 4$, so (2, 4) is on the line. In Fig. 1(a) we have drawn the various lines, and in Fig. 1(b) we have crossed out the appropriate regions to produce the graph of the system. The feasible set, as well as the equations of the various lines that make up its boundary, are shown in Fig. 2. From Fig. 2 we find the pairs of equations that determine each of the vertices A–D. This is the third step of the solution procedure.

$$A: \begin{cases} y = 3 \\ y = -x + 15 \end{cases} \qquad B: \begin{cases} y = 3 \\ y = -x + 27 \end{cases}$$

$$C: \begin{cases} y = -x + 27 \\ y = 2x \end{cases} \qquad D: \begin{cases} y = 2x \\ y = -x + 15. \end{cases}$$

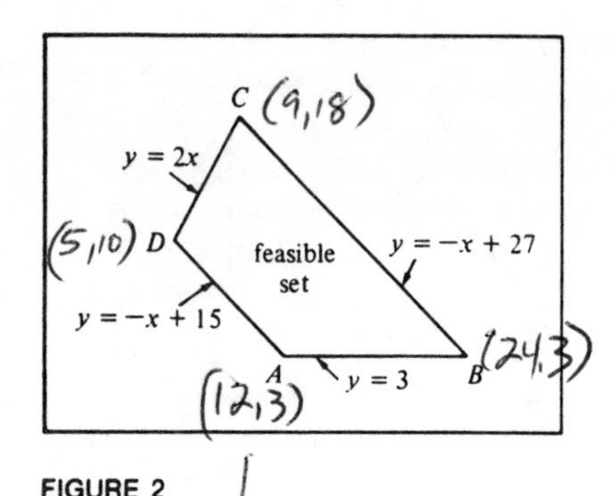

FIGURE 2

A and B are the easiest to determine. To find A, we must solve

$$3 = -x + 15$$
$$x = 12$$
$$y = 3$$
$$A = (12, 3).$$

Similarly, $B = (24, 3)$. To find C, we must solve

$$2x = -x + 27$$
$$3x = 27$$
$$x = 9$$
$$y = 2(9) = 18$$
$$C = (9, 18).$$

Similarly, $D = (5, 10)$.

Finally, we list the four vertices A, B, C, D and evaluate the objective function (2) at each one. The results are summarized in Table 3.

It is clear that the largest return occurs when $x = 5$, $y = 10$. In other words, $5 million should be invested in Treasury notes, $10 million in bonds, and $30 - (x + y) = 30 - (5 + 10) =$ $15 million in stocks.

Linear programming is of use not only in analyzing investments but in the fields of transportation and shipping. It is often used to plan routes, determine locations of warehouses, and develop efficient procedures for getting goods to people. Many linear programming problems of this variety can be formulated as *transportation problems.* A typical transportation problem involves determining the least-cost scheme for delivering a commodity stocked in a number of different warehouses to a number of different locations, say retail stores. Of course, in practical applications, it is necessary to consider problems involving perhaps dozens or even hundreds of warehouses, and possibly just as many delivery locations. For problems on such a grand scale, the methods developed so far are inadequate. For one thing, the number of variables required is usually more than two. We must wait until Chapter 3 for methods that apply to such problems. However, the next example gives an instance of a transportation problem which

TABLE 3

Vertex	*Return* $= 2.7 - .02x - .01y$
(5, 10)	$2.7 - .02(5) - .01(10) =$ $2.5 million
(9, 18)	$2.7 - .02(9) - .01(18) =$ $2.34 million
(24, 3)	$2.7 - .02(24) - .01(3) =$ $2.19 million
(12, 3)	$2.7 - .02(12) - .01(3) =$ $2.43 million

does not involve too many warehouses or too many delivery points. It gives the flavor of general transportation problems.

EXAMPLE 2 Suppose that a Maryland TV dealer has stores in Annapolis and Rockville and warehouses in College Park and Baltimore. The cost of shipping a set from College Park to Annapolis is \$6; from College Park to Rockville, \$3; from Baltimore to Annapolis, \$9; and from Baltimore to Rockville, \$5. Suppose that the Annapolis store orders 25 TV sets and the Rockville store 30. Suppose further that the College Park warehouse has a stock of 45 sets and the Baltimore warehouse 40. What is the most economical way to supply the requested TV sets to the two stores?

Solution The first step in solving a linear programming problem is to translate it into mathematical language. And the first part of this step is to organize the information given, preferably in the form of a chart. In this case, since the problem is geographic, we draw a schematic diagram, as in Fig. 3, which shows the flow of goods between warehouses and retail stores. By each route, we have written the cost. Below each warehouse we have written down its stock and below each retail store the number of TV sets it ordered.

Next, let us determine the variables. It appears initially that four variables are required, namely the number of TV sets to be shipped over each route. However, a closer look shows that only two variables are required. For if x denotes the number of TV sets to be shipped from College Park to Rockville, then since Rockville ordered 30 sets, the number shipped from Baltimore to Rockville is $30 - x$. Similarly, if y denotes the number of sets shipped from College Park to Annapolis, then the number shipped from Baltimore to Annapolis is $25 - y$. We have written the appropriate shipment sizes beside the various routes in Fig. 3.

As the third part of the translation process let us write down the restrictions on the variables. Basically, there are two kinds of restrictions: none of x, y, $30 - x$, $25 - y$ can be negative, and a warehouse cannot ship more TV sets than it has in stock. Referring to Fig. 3, we see that College Park ships $x + y$ sets, so that $x + y \leq 45$. Similarly, Baltimore ships $(30 - x) + (25 - y)$ sets, so that $(30 - x) + (25 - y) \leq 40$. Simplifying this inequality, we get

$$55 - x - y \leq 40$$

$$-x - y \leq -15$$

$$x + y \geq 15.$$

The inequality $30 - x \geq 0$ can be simplified to $x \leq 30$, and the inequality $25 -$

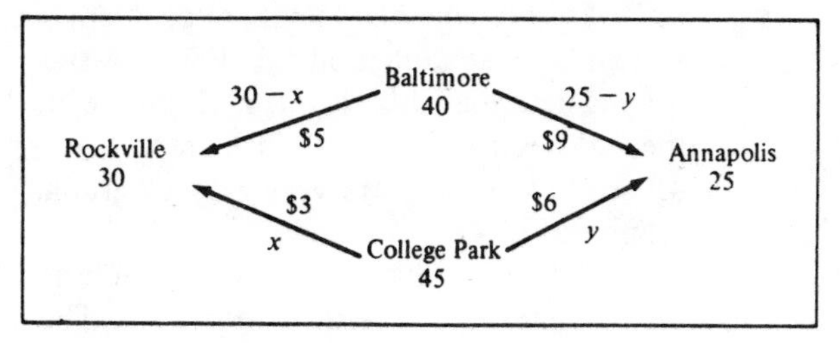

FIGURE 3

$y \geq 0$ can be written $y \leq 25$. So our restriction inequalities are these:

$$\begin{cases} x \geq 0, \quad y \geq 0 \\ x \leq 30, \quad y \leq 25 \\ x + y \geq 15 \\ x + y \leq 45. \end{cases} \tag{3}$$

The final step in the translation process is to form the objective function. In this problem we are attempting to minimize cost, so the objective function must express the cost in terms of x and y. Refer again to Fig. 3. There are x sets going from College Park to Rockville, and each costs \$3 to transport, so the cost of delivering these x sets is $3x$. Similarly, the costs of making the other deliveries are $6y$, $5(30 - x)$, and $9(25 - y)$, respectively. Thus the objective function is

$$\begin{aligned} [\text{cost}] &= 3x + 6y + 5(30 - x) + 9(25 - y) \\ &= 3x + 6y + 150 - 5x + 225 - 9y \\ &= 375 - 2x - 3y. \end{aligned} \tag{4}$$

FIGURE 4

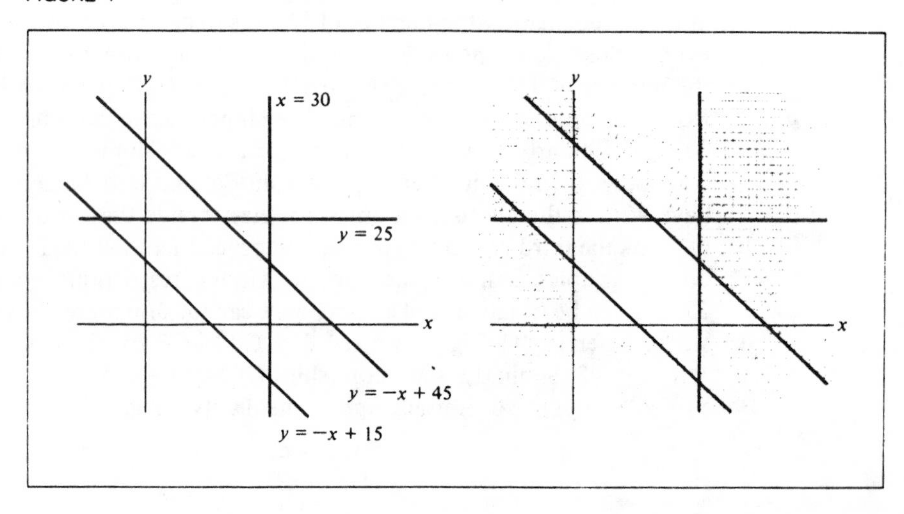

FIGURE 5

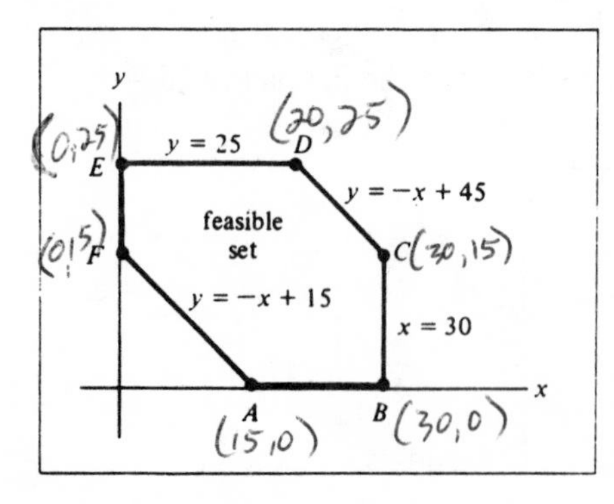

So the mathematical problem we must solve is: Find x and y that minimize the objective function (4) and satisfy the restrictions (3).

To solve the mathematical problem, we must graph the system of inequalities in (3). Four of the inequalities have graphs determined by horizontal and vertical lines. The only inequalities involving any work are $x + y \geq 15$ and $x + y \leq 45$. And even these are very easy to graph. The result is the graph in Fig. 4.

In Fig. 5 we have drawn the feasible set and have labeled each boundary line with its equation. The vertices A–F are now simple to determine. First, A and F are the intercepts of the line $y = -x + 15$. Therefore, $A = (15, 0)$ and $F = (0, 15)$. Since B is the x-intercept of the line $x = 30$, we have $B = (30, 0)$. Similarly, $E = (0, 25)$. Since C is on the line $x = 30$, its x-coordinate is 30. Its y-coordinate is $y = -30 + 45 = 15$, so $C = (30, 15)$. Similarly, since D has y-coordinate 25, its x-coordinate is given by $25 = -x + 45$ or $x = 20$. Thus $D = (20, 25)$.

We have listed in Table 4 the vertices A–F, as well as the cost corresponding to each one. The minimum cost occurs at the vertex (20, 25). So $x = 20$, $y = 25$ yields the minimum of the objective function. In other words, 20 TV sets should be shipped from College Park to Rockville and 25 from College Park to Annapolis, $30 - x = 10$ from Baltimore to Rockville, and $25 - y = 0$ from Baltimore to Annapolis. This solves our problem.

TABLE 4

Vertex	*Cost* $= 375 - 2x - 3y$
(0, 25)	300
(0, 15)	330
(15, 0)	345
(30, 0)	315
(30, 15)	270
(20, 25)	260

Remarks Concerning the Transportation Problem Note that the highest-cost route is the one from Baltimore to Annapolis. The solution we have obtained eliminates any shipments over this route. One might infer from this that one should always avoid the most expensive route. But this is not correct reasoning. To see why, reconsider Example 2, except change the cost of transporting a TV set from Baltimore to Annapolis from $9 to $7. The Baltimore-Annapolis route is still the most expensive. However, in this case the minimum cost is not obtained by eliminating the Baltimore-Annapolis route. For the revised problem, the linear inequalities stay the same. So the feasible set and the vertices remain the same. The only change is in the objective function, which now is given by

$$\begin{aligned}[\text{cost}] &= 3x + 6y + 5(30 - x) + 7(25 - y) \\ &= 325 - 2x - y.\end{aligned}$$

Therefore, the costs at the various vertices are as given in Table 5. So the minimum cost of $250 is achieved when $x = 30$, $y = 15$, $30 - x = 0$, and $25 - y = 10$. Note that 10 sets are being shipped from Baltimore to Annapolis, even though this is the most expensive route.

It is even possible for the cost function to be optimized simultaneously at two different vertices. For example, if the cost from Baltimore to Annapolis is $8 and all other data are the same as in Example 2, then the optimum cost is $260 and is

TABLE 5

Vertex	*Cost* = $325 - 2x - y$
(0, 25)	300
(0, 15)	310
(15, 0)	295
(30, 0)	265
(30, 15)	250
(20, 25)	260

achieved at both vertices (30, 15) and (20, 25). We leave the calculations to the reader.

Verification of the Fundamental Theorem

The fundamental theorem of linear programming asserts that the objective function assumes its optimum value at a vertex of the feasible set. Let us verify this fact. For simplicity, we give the argument only in a special case, namely for the furniture manufacturing problem. However, this is for convenience of exposition only. The same argument as given below may be used to prove the fundamental theorem in general. Our argument relies on the parallel property for straight lines, which asserts that parallel lines have the same slope.

EXAMPLE 3 Prove the fundamental theorem of linear programming in the special case of the furniture manufacturing problem.

Solution The profit derived from producing x chairs and y sofas is $80x + 70y$ dollars. Let us examine all those production schedules having a given profit. As an example, consider a profit of \$2800. Then x and y must satisfy $80x + 70y = 2800$. That is, (x, y) must lie on the line whose equation is $80x + 70y = 2800$, or in standard form, $y = -\frac{8}{7}x + 40$. The slope of this line is $-\frac{8}{7}$ and its y-intercept is (0, 40). We have drawn this line in Fig. 6(a), in which we have also drawn the feasible set for

FIGURE 6

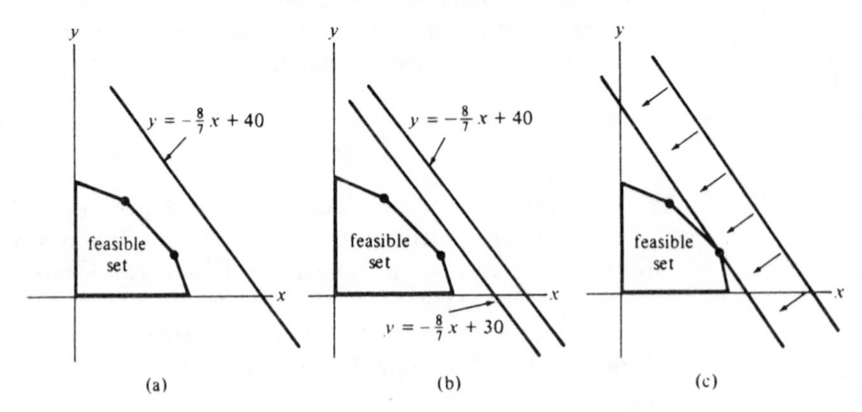

the furniture manufacturing problem. Note two fundamental facts: (1) Every production schedule on the line corresponds to a profit of $2800. (2) The line lies above the feasible set. In particular, no production schedule on the line satisfies all the restrictions of the problem. The difficulty is that $2800 is too high a profit for which to ask.

So now lower the profit, say, to $2100. In this case the production schedule (x, y) lies on the line $80x + 70y = 2100$, or in standard form, $y = -\frac{8}{7}x + 30$. This line is drawn in Fig. 6(b). Note that since both lines have slope $-\frac{8}{7}$, they are parallel by the parallel property. Actually, if we look at the production schedules yielding any fixed profit p, then they will lie along a line of slope $-\frac{8}{7}$, which is then parallel to the two lines already drawn. For if the production schedule (x, y) yields a profit p, then $80x + 70y = p$ or $y = -\frac{8}{7}x + p/70$. In other words, (x, y) lies on a line of slope $-\frac{8}{7}$ and y-intercept $(0, p/70)$. In particular, all the "lines of constant profit" are parallel to one another. So let us go back to the line of $2800 profit. It does not touch the feasible set. So now lower the profit and therefore translate the line downward parallel to itself. Next lower the profit until we first touch the feasible set. This line now touches the feasible set at a vertex [Fig. 6(c)]. And this vertex corresponds to the optimum production schedule, since any other point of the feasible set lies on a "line of constant profit" corresponding to an even lower profit. This shows why the fundamental theorem of linear programming is true.

PRACTICE PROBLEMS 3

Problems 1 and 2 refer to Example 1. Translate the statement into an inequality.

1. The amount to be invested in bonds is at most $5 million more than the amount to be invested in Treasury notes.
2. No more than $25 million should be invested in stocks and bonds.
3. Rework Example 1, assuming that the yield for Treasury notes goes up to 8%.
4. A linear programming problem has objective function: [cost] $= 5x + 10y$ which is to be minimized. Figure 7 shows the feasible set and the straight line of all combinations of x and y for which [cost] = $20.

   (a) Give the linear equation (in standard form) of the line of constant cost c.

   (b) As c increases, does the line of constant cost c move up or down?

   (c) By inspection, find the vertex of the feasible set that gives the optimum solution.

FIGURE 7

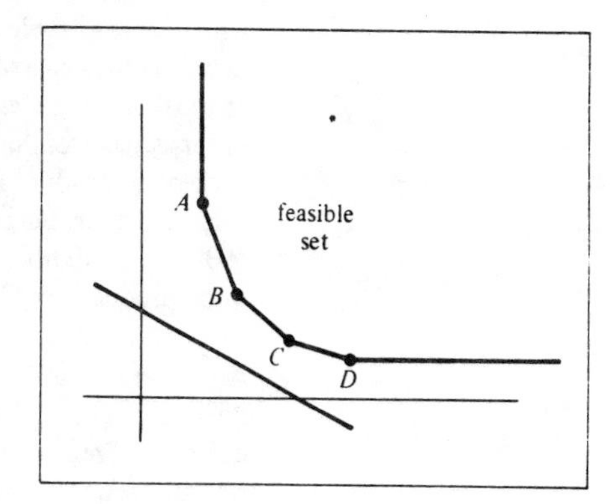

EXERCISES 3

1. Mr. Smith decides to feed his pet Doberman pinscher a combination of two dog foods. Each can of brand A contains 3 units of protein, 1 unit of carbohydrates, and 2 units of fat and costs 80 cents. Each can of brand B contains 1 unit of protein, 1 unit of carbohydrates, and 6 units of fat and costs 50 cents. Mr. Smith feels that each day his dog should have at least 6 units of protein, 4 units of carbohydrates, and 12 units of fat. How many cans of each dog food should he give to his dog each day in order to provide the minimum requirements at the least cost?

2. An oil company owns two refineries. Refinery I produces each day 100 barrels of high-grade oil, 200 barrels of medium-grade oil, and 300 barrels of low-grade oil and costs $10,000 to operate. Refinery II produces each day 200 barrels of high-grade, 100 barrels of medium-grade, and 200 barrels of low-grade oil and costs $9000 to operate. An order is received for 1000 barrels of high-grade oil, 1000 barrels of medium-grade oil, and 1800 barrels of low-grade oil. How many days should each refinery be operated in order to fill the order at the least cost?

3. A produce dealer in Florida ships oranges, grapefruits, and avocados to New York by truck. Each truckload consists of 100 crates, of which at least 20 crates must be oranges, at least 10 crates must be grapefruits, at least 30 crates must be avocados, and there must be at least as many crates of oranges as grapefruits. The profit per crate is $5 for oranges, $6 for grapefruits, and $4 for avocados. How many crates of each type should be shipped in order to maximize the profit? [*Hint:* Let x = number of crates of oranges, y = number of crates of grapefruit. Then $100 - x - y$ = number of crates of avocados.]

4. Mr. Jones has $9000 to invest in three types of stocks: low-risk, medium-risk, and high-risk. He invests according to three principles. The amount invested in low-risk stocks will be at most $1000 more than the amount invested in medium-risk stocks. At least $5000 will be invested in low- and medium-risk stocks. No more than $7000 will be invested in medium- and high-risk stocks. The expected yields are 6% for low-risk stock, 7% for medium-risk stocks, and 8% for high-risk stocks. How much money should Mr. Jones invest in each type of stock in order to maximize his total expected yield?

5. An automobile manufacturer has assembly plants in Detroit and Cleveland, each of which can assemble cars and trucks. The Detroit plant can assemble at most 800 vehicles in one day at a cost of $1200 per car and $2100 per truck. The Cleveland plant can assemble at most 500 vehicles in one day at a cost of $1000 per car and $2000 per truck. A rush order is received for 600 cars and 300 trucks. How many vehicles of each type should each plant produce in order to fill the order at the least cost? [*Hint:* Let x = number of cars to be produced in Detroit, y = number of trucks to be produced in Detroit, $600 - x$ = number of cars to be produced in Cleveland, $300 - y$ = number of trucks to be produced in Cleveland.]

6. A foreign car dealer with warehouses in New York and Baltimore receives orders from dealers in Philadelphia and Trenton. The dealer in Philadelphia needs 4 cars and the dealer in Trenton needs 7. The New York warehouse has 6 cars and the Baltimore warehouse has 8. The cost of shipping cars from Baltimore to Philadelphia is $120 per car,

from Baltimore to Trenton \$90 per car, from New York to Philadelphia \$100 per car, from New York to Trenton \$70 per car. Find the number of cars to be shipped from each warehouse to each dealer in order to minimize the shipping cost.

7. Figure 8(a) shows the feasible set of the nutrition problem of Section 2 and the straight line of all combinations of rice and soybeans for which the cost is 42 cents.

   (a) The objective function is $21x + 14y$. Give the linear equation (in standard form) of the line of constant cost c.

   (b) As c increases, does the line of constant cost move up or down?

   (c) By inspection, find the vertex of the feasible set that gives the optimum solution.

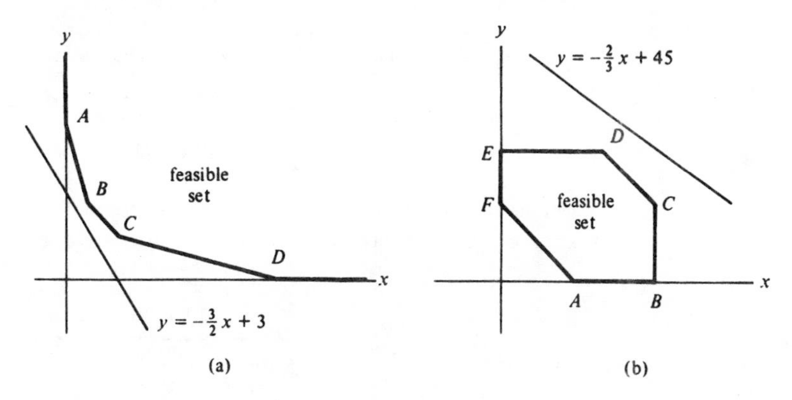

FIGURE 8

8. Figure 8(b) shows the feasible set of the transportation problem of Example 2 and the straight line of all combinations of shipments for which the transportation cost is \$240.

   (a) The objective function is [cost] $= 375 - 2x - 3y$. Give the linear equation (in standard form) of the line of constant cost c.

   (b) As c increases, does the line of constant cost move up or down?

   (c) By inspection, find the vertex of the feasible set that gives the optimum solution.

9. An oil refinery produces gasoline, jet fuel, and diesel fuel. The profits per gallon from the sale of these fuels are \$.15, \$.12, and \$.10, respectively. The refinery has a contract with an airline to deliver a minimum of 20,000 gallons per day of jet fuel and/or gasoline (or some of each). It has a contract with a trucking firm to deliver a minimum of 50,000 gallons per day of diesel fuel and/or gasoline (or some of each). The refinery can produce 100,000 gallons of fuel per day, distributed among the fuels in any fashion. It wishes to produce at least 5000 gallons per day of each fuel. How many gallons of each should be produced in order to maximize the profit?

10. Suppose that a price war reduces the profits of gasoline in Problem 9 to \$.05 per gallon and that the profits on jet fuel and diesel fuel are unchanged. How many gallons of each fuel should now be produced to maximize the profit?

SOLUTIONS TO PRACTICE PROBLEMS 3

1. Amount invested in bonds $= y$. Five million dollars more than the amount invested in Treasury notes is $x + 5$. Therefore, $y \le x + 5$.

2. Amount invested in stocks $= 30 - (x + y)$. Amount invested in bonds $= y$. Therefore,

$$30 - (x + y) + y \le 25$$

$$30 - x \le 25$$

$$x \ge 5.$$

3. The feasible set stays the same but the return becomes

$$\begin{aligned}[\text{return}] &= .08x + .08y + .09[30 - (x + y)] \\ &= .08x + .08y + 2.7 - .09x - 0.9y \\ &= 2.7 - .01x - .01y.\end{aligned}$$

When the return is evaluated at each of the vertices of the feasible set, the greatest return is achieved at two vertices. Either of these vertices yields an optimum solution.

	$2.7 - .01x - .01y$
(5, 10)	2.55
(12, 3)	2.55
(24, 3)	2.43
(9, 18)	2.43

4. (a) The values of x and y for which the cost is c dollars satisfy $5x + 10y = c$. The standard form of this linear equation is $y = -\frac{1}{2}x + c/10$.

   (b) The line $y = -\frac{1}{2}x + c/10$ has slope $-\frac{1}{2}$ and y-intercept $(0, c/10)$. As c increases, the slope stays the same, but the y-intercept moves up. Therefore, the line moves up.

   (c) The line of constant cost $20 does not contain any points of the feasible set, so such a low cost cannot be achieved. Increase the cost until the line of constant cost just touches the feasible set. As c increases, the line moves up (keeping the same slope) and first touches the feasible set at vertex C. Therefore, taking x and y to be the coordinates of C yields the minimum cost.

Chapter 6: CHECKLIST

- ☐ Objective function
- ☐ Fundamental theorem of linear programming
- ☐ Four-step procedure for solving linear programming problems

Chapter 6: SUPPLEMENTARY EXERCISES

1. Terrapin Airlines wants to fly 1400 members of a ski club to Colorado. The airline owns two types of planes. Type A can carry only 50 passengers, requires three stewards, and costs $14,000 for the trip. Type B can carry 300 passengers, requires four stewards, and costs $90,000 for the trip. If the airline must use at least as many type A planes as type B and has available only 42 stewards, how many planes of each type should be used to minimize the cost for the trip?

2. A nutritionist is designing a new breakfast cereal using wheat germ and enriched oat flour as the basic ingredients. Each ounce of wheat germ contains 2 milligrams of niacin, 3 milligrams of iron, .5 milligram of thiamin, and costs 3 cents. Each ounce of enriched oat flour contains 3 milligrams of niacin, 3 milligrams of iron, .25 milligram of thiamin, and costs 4 cents. The nutritionist wants the cereal to have at least 7 milligrams of niacin, 8 milligrams of iron, and 1 milligram of thiamin. How many ounces of wheat germ and how many ounces of enriched oat flour should be used in each serving in order to meet the nutritional requirements at the least cost?

3. An automobile manufacturer makes hardtops and sports cars. Each hardtop requires 8 man-hours to assemble, 2 man-hours to paint, 2 man-hours to upholster, and is sold for a profit of $90. Each sports car requires 18 man-hours to assemble, 2 man-hours to paint, 1 man-hour to upholster, and is sold for a profit of $100. During each day 360 man-hours are available to assemble, 50 man-hours to paint, and 40 man-hours to upholster automobiles. How many hardtops and sports cars should be produced each day in order to maximize the profit?

4. A confectioner makes two raisin-nut mixtures. A box of mixture A contains 6 ounces of peanuts, 1 ounce of raisins, 4 ounces of cashews, and sells for 50 cents. A box of mixture B contains 12 ounces of peanuts, 3 ounces of raisins, 2 ounces of cashews, and sells for 90 cents. He has available 5400 ounces of peanuts, 1200 ounces of raisins, and 2400 ounces of cashews. How many boxes of each mixture should he make in order to maximize revenue?

5. A textbook publisher puts out 72 new books each year, which are classified as elementary, intermediate, and advanced. The company's policy for new books is to publish at least four advanced books, at least three times as many elementary books as intermediate books, and at least twice as many intermediate books as advanced books. On the average, the annual profits are $8000 for each elementary book, $7000 for each intermediate book, and $1000 for each advanced book. How many new books of each type should be published in order to maximize the annual profit while conforming to company policy?

Chapter 3

The Simplex Method

In Chapter 2 we introduced a graphical method for solving linear programming problems. This method, although very simple, is of limited usefulness, since it applies only to problems which involve (or can be reduced to) two variables. On the other hand, linear programming applications in business and economics can involve dozens or even hundreds of variables. In this chapter we describe a method for handling such applications. This method, called the *simplex method* (or *simplex algorithm*), was developed by the mathematician George B. Dantzig in the late 1940s and today is the principal method used in solving complex linear programming problems. The simplex method can be used for problems in any number of variables and is easily adapted to computer calculations.

3.1 Slack Variables and the Simplex Tableau

In this section and the next, we explain how the simplex method can be used to solve linear programming problems such as the following:

PROBLEM A Maximize the objective function $3x + 4y$ subject to the constraints

$$\begin{cases} x + y \le 20 \\ x + 2y \le 25 \\ x \ge 0 \\ y \ge 0. \end{cases}$$

PROBLEM B Maximize the objective function $x + 2y + z$ subject to the constraints

$$\begin{cases} x - y + 2z \le 10 \\ 2x + y + 3z \le 12 \\ x \ge 0 \\ y \ge 0 \\ z \ge 0. \end{cases}$$

Each of these problems exhibits certain features that make it particularly convenient to work with.

1. The objective function is to be maximized.
2. Each variable is constrained to be ≥ 0.
3. All other constraints are of the form

$$[\text{linear polynomial}]^* \leq [\text{nonnegative constant}].$$

A linear programming problem satisfying these conditions is said to be in *standard form.* Our initial discussion of the simplex method will involve only such problems. Then, in Section 3, we will consider problems in nonstandard form.

The first step of the simplex method is to convert the given linear programming problem into a system of linear *equations.* To see how this is done, consider Problem A above. It specifies that the variables x and y are subject to the constraint

$$x + y \leq 20.$$

Let us introduce another variable, u, which turns the inequality into an equation:

$$x + y + u = 20.$$

The variable u "takes up the slack" between $x + y$ and 20 and is therefore called a *slack variable.* Moreover, since $x + y$ is at most 20, the variable u must be ≥ 0. In a similar way the constraint

$$x + 2y \leq 25$$

can be turned into the equation

$$x + 2y + v = 25,$$

where v is a slack variable and $v \geq 0$. Let us even turn our objective function $3x + 4y$ into an equation by introducing the new variable M defined by $M = 3x + 4y$. Then M is the variable we want to maximize. Moreover, it satisfies the equation

$$-3x - 4y + M = 0.$$

Thus, Problem A can be restated in terms of a system of linear equations as follows:

PROBLEM A′ Among all solutions of the system of linear equations

$$\begin{cases} x + y + u & = 20 \\ x + 2y + v & = 25 \\ -3x - 4y + M & = 0, \end{cases}$$

find one for which $x \geq 0$, $y \geq 0$, $u \geq 0$, $v \geq 0$, and for which M is as large as possible.

---

* A linear polynomial is an expression of the form $ax + by + cz + \cdots + dw$, where $a, b, c, \ldots, d$ are specific numbers and $x, y, z, \ldots, w$ are variables. Some examples are $2x - 3y + z$, $x + 2y + 3z - 4w$, and $-x + 3z - 2w$.

In a similar way, any linear programming problem in standard form can be reduced to that of determining a certain type of solution of a system of linear equations.

EXAMPLE 1 Formulate Problem B in terms of a system of linear equations.

Solution The two constraints $x - y + 2z \leq 10$, $2x + y + 3z \leq 12$ yield the equations

$$\begin{aligned} x - y + 2z + u \quad\quad &= 10 \\ 2x + y + 3z \quad\quad + v &= 12. \end{aligned}$$

The objective function yields the equation $M = x + 2y + z$—that is,

$$-x - 2y - z + M = 0.$$

So Problem B can be reformulated: Among all solutions of the system of linear equations

$$\begin{cases} x - y + 2z + u & = 10 \\ 2x + y + 3z \quad + v & = 12 \\ -x - 2y - z \quad\quad + M & = 0, \end{cases}$$

find one for which $x \geq 0, y \geq 0, z \geq 0, u \geq 0, v \geq 0$, and M is as large as possible.

We shall now discuss a scheme for solving systems of equations like those just encountered. For the moment, we will not worry about maximizing M or keeping the variables ≥ 0. Rather, let us concentrate on a particular method for determining solutions. In order to be concrete, consider the system of linear equations from Problem A′:

$$\begin{cases} x + y + u & = 20 \\ x + 2y \quad + v & = 25 \\ -3x - 4y \quad\quad + M & = 0, \end{cases} \tag{1}$$

This system has an infinite number of solutions. In fact, we can solve the equations for u, v, M in terms of x and y:

$$\begin{aligned} u &= 20 - x - y \\ v &= 25 - x - 2y \\ M &= \quad\quad 3x + 4y. \end{aligned}$$

Given any values of x and y, we can determine corresponding values for u, v, M. For example, if $x = 0$ and $y = 0$, then

$$\begin{aligned} u &= 20 - 0 - 0 = 20 \\ v &= 25 - 0 - 2 \cdot 0 = 25 \\ M &= \quad\quad 3 \cdot 0 + 4 \cdot 0 = 0. \end{aligned}$$

Therefore, $x = 0$, $y = 0$, $u = 20$, $v = 25$, $M = 0$ is a specific solution of the system. Note that these values for u, v, M are precisely the numbers that appear to the right of the equality signs in our system of linear equations. Therefore, this particular solution could have been read off directly from the system (1) without any computation. This method of generating solutions is used in the simplex method, so let us explore further the special properties of the system which allowed us to read off a specific solution so easily.

Note that the system of linear equations has five variables: x, y, u, v, M. These variables may be divided into two groups. Group I consists of those which were set equal to 0, namely x and y. Group II consists of those whose particular values were read off from the right-hand sides of the equations, namely u, v, and M. Note also that the system has a special form which allows the particular values of the group II variables to be read off: each of the equations involves exactly one of the group II variables, and these variables always appear with coefficient 1. Thus, for example, the first equation involves the group II variable u:

$$x + y + u = 20.$$

Therefore, when all group I variables (x and y) are set equal to 0, only the term u remains on the left, and the particular value of u can then be read off from the right-hand side.

The special form of the system can best be described in matrix form. Write the system in the usual way as a matrix, but add column headings corresponding to the variables:

$$\begin{array}{c} \begin{matrix} \;\; x & \;\; y & \;\; u & \;\; v & \;\; M & \quad \end{matrix} \\ \left[\begin{array}{rrrrr|r} 1 & 1 & 1 & 0 & 0 & 20 \\ 1 & 2 & 0 & 1 & 0 & 25 \\ -3 & -4 & 0 & 0 & 1 & 0 \end{array}\right] \end{array}.$$

Note closely the columns corresponding to the group II variables u, v, M.

$$\begin{array}{c} \begin{matrix} \;\; x & \;\; y & \;\; u & \;\; v & \;\; M & \quad \end{matrix} \\ \left[\begin{array}{rrrrr|r} 1 & 1 & 1 & 0 & 0 & 20 \\ 1 & 2 & 0 & 1 & 0 & 25 \\ -3 & -4 & 0 & 0 & 1 & 0 \end{array}\right] \end{array}.$$

The presence of these columns gives the system the special form discussed above. Indeed, the u column asserts that u appears only in the first equation and is coefficient there is 1, and similarly for the v and M columns.

The property of allowing us to read off a particular solution from the right-hand column is shared by all linear systems whose matrices contain the columns

$$\begin{matrix} 1 & 0 & 0 & \cdots & 0 \\ 0 & 1 & 0 & \cdots & 0 \\ 0 & 0 & 1 & \cdots & 0 \\ \vdots & \vdots & \vdots & & \vdots \\ 0 & 0 & 0 & & 1. \end{matrix}$$

(These columns need not appear in exactly the order shown.) The variables corresponding to these columns are called the group II variables. The group I variables consist of all the others. To get one particular solution to the system, set all the group I variables equal to zero and read off the values of the group II variables from the right-hand side of the system. This procedure is illustrated in the following example.

EXAMPLE 2 Determine by inspection one set of solutions to each of these systems of linear equations:

(a) $$\begin{cases} x - 5y + u = 3 \\ -2x + 8y + v = 11 \\ -\frac{1}{2}x + M = 0 \end{cases}$$

(b) $$\begin{cases} -y + 2u + v = 12 \\ x + \frac{1}{2}y - 6u = -1 \\ 3y + 8u + M = 4 \end{cases}$$

Solution (a) The matrix of the system is

$$\begin{array}{ccccc|c} x & y & u & v & M & \\ 1 & -5 & 1 & 0 & 0 & 3 \\ -2 & 8 & 0 & 1 & 0 & 11 \\ -\frac{1}{2} & 0 & 0 & 0 & 1 & 0 \end{array}.$$

We look for each variable whose column contains one entry of 1 and all the other entries 0.

$$\begin{array}{ccccc|c} x & y & u & v & M & \\ 1 & -5 & 1 & 0 & 0 & 3 \\ -2 & 8 & 0 & 1 & 0 & 11 \\ -\frac{1}{2} & 0 & 0 & 0 & 1 & 0 \end{array}.$$

The group II variables should be u, v, M, with x, y as the group I variables. Set all group I variables equal to 0. The corresponding values of the group II variables may then be read off the last column: $u = 3$, $v = 11$, $M = 0$. So one solution of the system is

$$x = 0, \quad y = 0, \quad u = 3, \quad v = 11, \quad M = 0.$$

(b) The matrix of the system is

$$\begin{array}{ccccc|c} x & y & u & v & M & \\ 0 & -1 & 2 & 1 & 0 & 12 \\ 1 & \frac{1}{2} & -6 & 0 & 0 & -1 \\ 0 & 3 & 8 & 0 & 1 & 4 \end{array}.$$

The shaded columns show that the group II variables should be v, x, M, with y, u, as the group I variables. So the corresponding solution is

$$x = -1, \quad y = 0, \quad u = 0, \quad v = 12, \quad M = 4.$$

A *simplex tableau* is a matrix (corresponding to a linear system) in which each of the columns

$$\begin{matrix} 1 & 0 & \cdots & 0 \\ 0 & 1 & \cdots & 0 \\ \vdots & \vdots & & \vdots \\ 0 & 0 & & 1 \end{matrix}$$

is present exactly once (in some order) to the left of the vertical line. We have seen how to construct a simplex tableau corresponding to a linear programming problem in standard form. From this initial simplex tableau we can read off one particular solution of the linear system by using the method described above. This particular solution may or may not correspond to the solution of the original optimization problem. If it does not, we replace the initial tableau with another one whose corresponding solution is "closer" to the optimum. How do we replace the initial simplex tableau with another? Just pivot it about a nonzero entry! Indeed, one of the key reasons the simplex method works is that pivoting transforms one simplex tableau into another. Note also that since pivoting consists of elementary row operations, the solution corresponding to a transformed tableau is a solution of the original linear system. The next example illustrates how pivoting transforms a tableau into another one.

EXAMPLE 3 Consider the simplex tableau obtained from Problem A:

$$\begin{array}{ccccc} x & y & u & v & M & \\ \end{array}$$
$$\left[\begin{array}{ccccc|c} 1 & 1 & 1 & 0 & 0 & 20 \\ 1 & \textcircled{2} & 0 & 1 & 0 & 25 \\ -3 & -4 & 0 & 0 & 1 & 0 \end{array}\right].$$

(a) Pivot this tableau around the entry 2.

(b) Calculate the particular solution corresponding to the transformed tableau which results from setting the new group I variables equal to 0.

Solution (a) The first step in pivoting is to replace the pivot element 2 by a 1. To do this, multiply the second row of the tableau by $\frac{1}{2}$ to get

$$\left[\begin{array}{ccccc|c} 1 & 1 & 1 & 0 & 0 & 20 \\ \frac{1}{2} & 1 & 0 & \frac{1}{2} & 0 & \frac{25}{2} \\ -3 & -4 & 0 & 0 & 1 & 0 \end{array}\right].$$

Next, we must replace all nonpivot elements in the second column by zeros. Do this by adding to the first row (-1) times the second row:

$$\left[\begin{array}{ccccc|c} \frac{1}{2} & 0 & 1 & -\frac{1}{2} & 0 & \frac{15}{2} \\ \frac{1}{2} & 1 & 0 & \frac{1}{2} & 0 & \frac{25}{2} \\ -3 & -4 & 0 & 0 & 1 & 0 \end{array}\right],$$

and by adding to the third row 4 times the second row:

$$\begin{array}{ccccc|c} x & y & u & v & M & \\ \end{array}$$
$$\left[\begin{array}{ccccc|c} \frac{1}{2} & 0 & 1 & -\frac{1}{2} & 0 & \frac{15}{2} \\ \frac{1}{2} & 1 & 0 & \frac{1}{2} & 0 & \frac{25}{2} \\ -1 & 0 & 0 & 2 & 1 & 50 \end{array}\right].$$

Note that we indeed get a new simplex tableau. The new group II variables are now u, y, M. The new group I variables are x, v.

$$\begin{array}{ccccc|c} x & y & u & v & M & \\ \end{array}$$
$$\left[\begin{array}{ccccc|c} \frac{1}{2} & 0 & 1 & -\frac{1}{2} & 0 & \frac{15}{2} \\ \frac{1}{2} & 1 & 0 & \frac{1}{2} & 0 & \frac{25}{2} \\ -1 & 0 & 0 & 2 & 1 & 50 \end{array}\right].$$

(b) Set the group I variables equal to 0:

$$x = 0, \qquad v = 0.$$

Read off the particular values of the group II variables from the right-hand column:

$$u = \tfrac{15}{2}, \qquad y = \tfrac{25}{2}, \qquad M = 50.$$

So the particular solution corresponding to the transformed tableau is

$$x = 0, \qquad y = \tfrac{25}{2}, \qquad u = \tfrac{15}{2}, \qquad v = 0, \qquad M = 50.$$

PRACTICE PROBLEMS 1

1. Determine by inspection a particular solution of the following system of linear equations:

$$\begin{cases} x + 2y + 3u = 6 \\ y + v = 4 \\ 5y + 2u + M = 0. \end{cases}$$

2. Pivot the simplex tableau about the circled element:

$$\left[\begin{array}{ccccc|c} 2 & 4 & 1 & 0 & 0 & 6 \\ 3 & \textcircled{1} & 0 & 1 & 0 & 0 \\ 1 & 1 & 0 & 0 & 1 & 1 \end{array}\right].$$

EXERCISES 1

For each of the following linear programming problems, determine the corresponding linear system and restate the linear programming problem in terms of the linear system.

1. Maximize $8x + 13y$ subject to the constraints

$$\begin{cases} 20x + 30y \le 3500 \\ 50x + 10y \le 5000 \\ x \ge 0 \\ y \ge 0. \end{cases}$$

2. Maximize $x + 15y$ subject to the constraints

$$\begin{cases} 3x + 2y \leq 10 \\ x \quad\quad \leq 15 \\ \quad\quad y \leq 3 \\ x + \ y \leq 5 \\ x \geq 0 \\ y \geq 0. \end{cases}$$

3. Maximize $x + 2y - 3z$ subject to the constraints

$$\begin{cases} x + \quad y + z \leq 100 \\ 3x \quad\quad + z \leq 200 \\ 5x + 10y \quad\quad \leq 100 \\ x \geq 0 \\ y \geq 0 \\ z \geq 0. \end{cases}$$

4. Maximize $2x + y + 50$ subject to the constraints

$$\begin{cases} x + 3y \leq 24 \\ \quad\quad y \leq 5 \\ x + 7y \leq 10 \\ x \geq 0 \\ y \geq 0. \end{cases}$$

5-8. For each of the linear programming problems in Exercises 1–4:

(a) Set up the simplex tableau.

(b) Determine the particular solution corresponding to the tableau.

Find the particular solutions corresponding to these tableaux.

9.
$$\begin{array}{ccccc|c} x & y & u & v & M & \\ 0 & 2 & 1 & 0 & 0 & 10 \\ 1 & 3 & 0 & 12 & 0 & 15 \\ 0 & -1 & 0 & 17 & 1 & 20 \end{array}$$

10.
$$\begin{array}{ccccc|c} x & y & u & v & M & \\ 1 & 0 & 3 & 11 & 0 & 6 \\ 0 & 1 & 10 & 17 & 0 & 16 \\ 0 & 0 & 5 & -1 & 1 & 3 \end{array}$$

11.
$$\begin{array}{ccccccc|c} x & y & z & u & v & w & M & \\ 0 & 3 & 1 & 0 & 1 & 15 & 0 & 15 \\ 1 & -1 & 0 & 0 & 2 & -5 & 0 & 10 \\ 0 & 2 & 0 & 1 & -5 & 4 & 0 & 23 \\ 0 & 11 & 0 & 0 & 11 & 6 & 1 & -11 \end{array}$$

12. $$\begin{array}{c} \begin{array}{ccccccc|c} x & y & z & u & v & w & M & \end{array} \\ \left[\begin{array}{ccccccc|c} 6 & 0 & 1 & 0 & 5 & -1 & 0 & \frac{1}{4} \\ 5 & 1 & 0 & 0 & 3 & \frac{1}{3} & 0 & 100 \\ 4 & 0 & 0 & 1 & 8 & \frac{1}{2} & 0 & 11 \\ 2 & 0 & 0 & 0 & 6 & \frac{1}{7} & 1 & -\frac{1}{2} \end{array}\right] \end{array}$$

13. Pivot the simplex tableau

$$\begin{array}{c} \begin{array}{ccccc|c} x & y & u & v & M & \end{array} \\ \left[\begin{array}{ccccc|c} 2 & 3 & 1 & 0 & 0 & 12 \\ 1 & 1 & 0 & 1 & 0 & 10 \\ -10 & -20 & 0 & 0 & 1 & 0 \end{array}\right] \end{array}$$

about the indicated element and compute the particular solution corresponding to the new tableau.

(a) 2

(b) 3

(c) 1 (second row, first column)

(d) 1 (second row, second column).

14. Pivot the simplex tableau

$$\begin{array}{c} \begin{array}{ccccc|c} x & y & u & v & M & \end{array} \\ \left[\begin{array}{ccccc|c} 5 & 4 & 1 & 0 & 0 & 100 \\ 10 & 6 & 0 & 1 & 0 & 1200 \\ -1 & 2 & 0 & 0 & 1 & 0 \end{array}\right] \end{array}$$

about the indicated element and compute the solution corresponding to the new tableau.

(a) 5 (b) 4 (c) 10 (d) 6.

15. Determine which of the pivot operations in Exercise 13 increases M the most.

16. Determine which of the pivot operations in Exercise 14 increases M the most.

SOLUTIONS TO PRACTICE PROBLEMS 1

1. The matrix of the system is

$$\begin{array}{c} \begin{array}{ccccc|c} x & y & u & v & M & \end{array} \\ \left[\begin{array}{ccccc|c} 1 & 2 & 3 & 0 & 0 & 6 \\ 0 & 1 & 0 & 1 & 0 & 4 \\ 0 & 5 & 2 & 0 & 1 & 0 \end{array}\right] \end{array},$$

from which we see that the group II variables are x, v, M, and the group I variables y, u. To obtain a solution, we set the group I variables equal to 0. We obtain from the first equation that $x = 6$, from the second that $v = 4$, and from the third that $M = 0$. Thus a solution of the system is $x = 6$, $y = 0$, $u = 0$, $v = 4$, $M = 0$.

2. We must use elementary row operations to transform the second column into

$$\begin{bmatrix} 0 \\ 1 \\ 0 \end{bmatrix}.$$

$$\begin{bmatrix} 2 & 4 & 1 & 0 & 0 & | & 6 \\ 3 & 1 & 0 & 1 & 0 & | & 0 \\ 1 & 1 & 0 & 0 & 1 & | & 1 \end{bmatrix} \xrightarrow{[1] + (-4)[2]} \begin{bmatrix} -10 & 0 & 1 & -4 & 0 & | & 6 \\ 3 & 1 & 0 & 1 & 0 & | & 0 \\ 1 & 1 & 0 & 0 & 1 & | & 1 \end{bmatrix}$$

$$\xrightarrow{[3] + (-1)[2]} \begin{bmatrix} -10 & 0 & 1 & -4 & 0 & | & 6 \\ 3 & 1 & 0 & 1 & 0 & | & 0 \\ -2 & 0 & 0 & -1 & 1 & | & 1 \end{bmatrix}.$$

3.2 The Simplex Method, I—Maximum Problems

We can now describe the simplex method for solving linear programming problems. The procedure will be illustrated as we solve Problem A of Section 1. Recall that we must maximize the objective function $3x + 4y$ subject to the constraints

$$\begin{cases} x + \ \ y \leq 20 \\ x + 2y \leq 25 \\ x \geq 0, \quad y \geq 0. \end{cases}$$

> *Step 1* Introduce slack variables and state the problem in terms of a system of linear equations.

We carried out this step in the preceding section. The result was the following restatement of the problem.

PROBLEM A′ Among the solutions of the system of linear equations

$$\begin{cases} x + \ \ y + u \qquad\qquad = 20 \\ x + 2y \qquad + v \qquad = 25 \\ -3x - 4y \qquad\quad + M = 0, \end{cases}$$

find one for which $x \geq 0$, $y \geq 0$, $u \geq 0$, $v \geq 0$, and for which M is as large as possible.

> *Step 2* Construct the simplex tableau corresponding to the linear system.

This step was alo carried out in Section 1. The tableau is

$$\begin{array}{c} \\ u \\ v \\ M \end{array} \begin{array}{c} \begin{array}{ccccc} x & y & u & v & M \end{array} \\ \left[\begin{array}{ccccc|c} 1 & 1 & 1 & 0 & 0 & 20 \\ 1 & 2 & 0 & 1 & 0 & 25 \\ \hline -3 & -4 & 0 & 0 & 1 & 0 \end{array}\right] \end{array}. \qquad (1)$$

Note that we have made two additions to the previously found tableau. First, we have separated the last row from the others by means of a horizontal line. This is because the last row, which corresponds to the objective function in the original problem, will play a special role in what follows. The second addition is that we have labeled each row with one of the group II variables—namely, the variable whose value is determined by the row. Thus, for example, the first row gives the particular value of u, which is 20, so the row is labeled with a u. We will find these labels convenient.

Corresponding to this tableau, there is a particular solution to the linear system, namely the one obtained by setting all group I variables equal to 0. Reading the values of the group II variables from the last column, we obtain

$$x = 0, \qquad y = 0, \qquad u = 20, \qquad v = 25, \qquad M = 0.$$

Our object is to make M as large as possible. How can the value of M be increased? Look at the equation corresponding to the last row of the tableau. It reads

$$-3x - 4y + M = 0.$$

Note that two of the coefficients, -3 and -4, are negative. Or, what amounts to the same thing, if we solve for M and get

$$M = 3x + 4y,$$

then the coefficients on the right-hand side are *positive*. This fact is significant. It says that M can be increased by increasing either the value of x or the value of y. A unit change in x will increase M by 3 units, whereas a unit change in y will increase M by 4 units. And since we wish to increase M by as much as possible, it is reasonable to attempt to increase the value of y. Let us indicate this by drawing an arrow pointing to the y column of the tableau:

$$\begin{array}{c} \\ u \\ v \\ M \\ \\ \end{array} \begin{array}{c} \begin{array}{cccccc} x & y & u & v & M & \end{array} \\ \left[\begin{array}{ccccc|c} 1 & 1 & 1 & 0 & 0 & 20 \\ 1 & 2 & 0 & 1 & 0 & 25 \\ \hline -3 & -4 & 0 & 0 & 1 & 0 \end{array}\right] \\ \begin{array}{cccccc} & \uparrow & & & & \end{array} \end{array}.$$

In order to increase y (from its present value, zero), we will pivot about one of the entries (above the horizontal line) in the y column. In this way, y will become a group II variable and hence will not necessarily be zero in our next particular solution. But which entry should we pivot around? To find out, let us experiment. The results from pivoting about the 1 and the 2 in the y column are, respectively,

$$\begin{array}{c} \\ y \\ v \\ M \end{array} \begin{array}{c} \begin{array}{cccccc} x & y & u & v & M & \end{array} \\ \left[\begin{array}{ccccc|c} 1 & 1 & 1 & 0 & 0 & 20 \\ -1 & 0 & -2 & 1 & 0 & -15 \\ \hline 1 & 0 & 4 & 0 & 1 & 80 \end{array}\right] \end{array}$$

Pivot about 1

$$
\begin{array}{c}
 \\ u \\ y \\ M
\end{array}
\begin{array}{c}
\begin{array}{cccccc} x & y & u & v & M & \end{array} \\
\left[\begin{array}{rrrrr|r}
\frac{1}{2} & 0 & 1 & -\frac{1}{2} & 0 & \frac{15}{2} \\
\frac{1}{2} & 1 & 0 & \frac{1}{2} & 0 & \frac{25}{2} \\
\hline
-1 & 0 & 0 & 2 & 1 & 50
\end{array}\right].
\end{array}
$$

Pivot about 2

Note that the labels on the rows have *changed* because the group II variables are now *different*. The solutions corresponding to these tableaux are, respectively,

$$x = 0, \quad y = 20, \quad u = 0, \quad v = -15, \quad M = 80$$

and

$$x = 0, \quad y = \tfrac{25}{2}, \quad u = \tfrac{15}{2}, \quad v = 0, \quad M = 50.$$

The first solution violates the requirement that all variables be ≥ 0. Thus we use the second solution, which arose from pivoting about 2. Using this solution, we have increased the value of M to 50 and have replaced our original tableau by

$$
\begin{array}{c}
 \\ u \\ y \\ M
\end{array}
\begin{array}{c}
\begin{array}{cccccc} x & y & u & v & M & \end{array} \\
\left[\begin{array}{rrrrr|r}
\frac{1}{2} & 0 & 1 & -\frac{1}{2} & 0 & \frac{15}{2} \\
\frac{1}{2} & 1 & 0 & \frac{1}{2} & 0 & \frac{25}{2} \\
\hline
-1 & 0 & 0 & 2 & 1 & 50
\end{array}\right].
\end{array}
$$

Can M be increased further? To answer this question, look at the last row of the tableau, corresponding to the equation

$$-x + 2v + M = 50.$$

There is a negative coefficient for the variable x in this equation. Correspondingly, when the equation is solved for M, there is a positive coefficient for x:

$$M = 50 + x - 2v.$$

Now it is clear that we should try to increase x. So we pivot about one of the entries in the x column. A quick calculation shows that pivoting about the second entry leads to a solution having some negative values. Therefore, we pivot about the top entry. The result is

$$
\begin{array}{c}
 \\ x \\ y \\ M
\end{array}
\begin{array}{c}
\begin{array}{cccccc} x & y & u & v & M & \end{array} \\
\left[\begin{array}{rrrrr|r}
1 & 0 & 2 & -1 & 0 & 15 \\
0 & 1 & -1 & 1 & 0 & 5 \\
\hline
0 & 0 & 2 & 1 & 1 & 65
\end{array}\right].
\end{array}
$$

The corresponding solution is

$$x = 15, \quad y = 5, \quad u = 0, \quad v = 0, \quad M = 65.$$

Note that with this second pivot operation we have increased the value of M from 50 to 65.

Can we increase M still further? Let us reason as before. Use the last row of the current tableau to write M in terms of the other variables:

$$2u + v + M = 65, \qquad M = 65 - 2u - v.$$

Note, however, that in contrast to the previous expressions for M, this one has *no positive coefficients*. And since u and v are ≥ 0, this means that M can be *at most* 65. But M is already 65. So M cannot be increased further. Thus we have shown that the maximum value of M is 65, and this occurs when $x = 15$, $y = 5$. This solves the original linear programming problem.

Based on the discussion above, we can state several general principles. First of all, the following criterion determines when a simplex tableau yields a maximum.

Condition for a Maximum The particular solution derived from a simplex tableau is a maximum if and only if the bottom row contains no negative entries except perhaps the entry in the last column.*

We saw this condition illustrated in the example above. Each of the first two tableaux had negative entries in the last row, and, as we showed, their corresponding solutions were not maxima. However, the third tableau, with no negative entries in the last row, did yield a maximum.

The crucial point of the simplex method is the correct choice of a pivot element. In the example above we decided to choose a pivot element from the column corresponding to the most negative entry in the last row. It can be proved that this is the proper choice in general; that is, we have the following rule:

Choosing the Pivot Column The pivot element should be chosen from that column to the left of the vertical line which has the most negative entry in the last row.†

Choosing the correct pivot element from the designated column is somewhat more complicated. Our approach above was to calculate the tableau associated with each element and observe that only one corresponded to a solution with nonnegative elements. However, there is a simpler way to make the choice. As an illustration, let us reconsider tableau (1). We have already decided to pivot around some entry in the second column. For each *positive* entry in the pivot column we compute a ratio: the corresponding entry in the right-hand column divided by the entry in the pivot column. So for example, for the first entry the ratio is $\frac{20}{1}$ and

*In Section 3 we shall encounter maximum problems whose final tableaux have a negative number in the lower right-hand corner.

† In case two or more columns are tied for the honor of being pivot column, an arbitrary choice among them may be made.

for the second $\frac{25}{2}$. We write these ratios to the right of the matrix as follows:

$$\begin{array}{c} \\ u \\ v \\ M \end{array} \begin{array}{c} \begin{array}{ccccc} x & y & u & v & M \end{array} \\ \left[\begin{array}{ccccc|c} 1 & 1 & 1 & 0 & 0 & 20 \\ 1 & 2 & 0 & 1 & 0 & 25 \\ \hline -3 & -4 & 0 & 0 & 1 & 0 \end{array}\right] \end{array} \begin{array}{c} \\ 20/1 \\ 25/2 \\ \\ \end{array}$$

It is possible to prove the following rule, which allows us to determine the pivot element from the above display:

> *Choosing the Pivot Element* For each positive entry of the pivot column, compute the appropriate ratio. Choose as pivot element the one corresponding to the least (nonnegative) ratio.

For instance, consider the choice of pivot element in the example above. The least among the ratios is $\frac{25}{2}$. So we choose 2 as the pivot element.

At first, this method for choosing the pivot element might seem very odd. However, it is just a way of guaranteeing that the last column of the new tableau will have entries ≥ 0. And that is just the basis on which we chose the pivot element earlier. To obtain further insight, let us analyze the example above yet further.

Suppose that we pivot our tableau about the 1 in column 2. The first step in pivoting is to divide the pivot row by the pivot element (in this case 1). This gives the array

$$\left[\begin{array}{ccccc|c} 1 & 1 & 1 & 0 & 0 & \frac{20}{1} \\ 1 & 2 & 0 & 1 & 0 & 25 \\ \hline -3 & -4 & 0 & 0 & 1 & 0 \end{array}\right],$$

where we have written $\frac{20}{1}$ rather than 20 to emphasize that we have divided by the pivot element. The next step in the pivot procedure is to replace the second row by the second row plus (-2) times the first row. The result is

$$\left[\begin{array}{ccccc|c} 1 & 1 & 1 & 0 & 0 & \frac{20}{1} \\ -1 & 0 & -2 & 1 & 0 & 25 + (-2)\frac{20}{1} \\ \hline -3 & -4 & 0 & 0 & 1 & 0 \end{array}\right].$$

The final step of the pivot process is to replace the third row by $[3] + 4[1]$, to obtain

$$\left[\begin{array}{ccccc|c} 1 & 1 & 1 & 0 & 0 & \frac{20}{1} \\ -1 & 0 & -2 & 1 & 0 & 25 + (-2)\frac{20}{1} \\ \hline 1 & 0 & 4 & 0 & 1 & 80 \end{array}\right].$$

Notice that the two upper entries in the last column may be written

$$\frac{20}{1}, \qquad 2 \cdot \left(\frac{25}{2} - \frac{20}{1}\right).$$

Notice the difference of the ratios appearing in the second entry! If we similarly pivot about the 2 in the second column of the original tableau, we obtain

$$\left[\begin{array}{ccccc|c} \frac{1}{2} & 0 & 1 & -\frac{1}{2} & 0 & 20 + (-1)\frac{25}{2} \\ \frac{1}{2} & 1 & 0 & \frac{1}{2} & 0 & \frac{25}{2} \\ \hline -1 & 0 & 0 & 2 & 1 & 50 \end{array}\right].$$

The upper entries in the last column are

$$\frac{20}{1} - \frac{25}{2}, \quad \frac{25}{2}.$$

Again notice the difference of the ratios, this time the reverse of the previous difference. Note that when we pivot about 1, we subtract the ratio 20/1, whereas if we pivot about 2, we subtract the ratio 25/2. In order to arrive at nonnegative entries in the last column, we should subtract off as little as possible. That is, we should pivot about the entry corresponding to the smallest ratio. This is the rationale governing our choice of pivot element!

Now that we have assembled all the components of the simplex method, we can summarized it as follows:

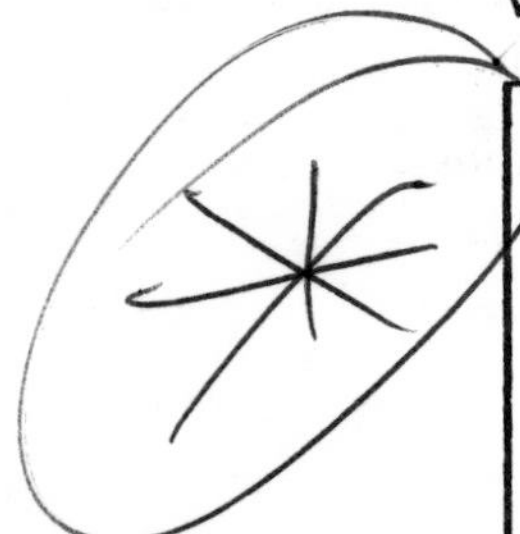

The Simplex Method for Problems in Standard Form

1. Introduce slack variables and state the problem in terms of a system of linear equations.
2. Construct the simplex tableau corresponding to the system.
3. Determine if the left part of the bottom row contains negative entries. If none are present, the solution corresponding to the tableau yields a maximum and the problem is solved.
4. If the left part of the bottom row contains negative entries, construct a new simplex tableau.
   (a) Choose the pivot column by inspecting the entries of the last row of the current tableau, excluding the right-hand entry. The pivot column is the one containing the most negative of these entries.
   (b) Choose the pivot element by computing ratios associated with the positive entries of the pivot column. The pivot element is the one corresponding to the smallest (nonnegative) ratio.
   (c) Construct the new simplex tableau by pivoting around the selected element.
5. Return to step 3. Steps 3 and 4 are repeated as many times as necessary to find a maximum.

Let us now work some problems to see how this method is applied.

EXAMPLE 1 Maximize the objective function $10x + y$ subject to the constraints

$$\begin{cases} x + 2y \le 10 \\ 3x + 4y \le 6 \\ x \ge 0, \quad y \ge 0. \end{cases}$$

Solution The corresponding system of linear equations with slack variables is

$$\begin{cases} x + 2y + u = 10 \\ 3x + 4y + v = 6 \\ -10x - y + M = 0, \end{cases}$$

and we must find that solution of the system for which $x \geq 0, y \geq 0, u \geq 0, v \geq 0$, and M is as large as possible. Here is the initial simplex tableau:

$$\begin{array}{c} \\ u \\ v \\ M \end{array} \begin{array}{c} \begin{array}{ccccc|c} x & y & u & v & M & \end{array} \\ \left[\begin{array}{ccccc|c} 1 & 2 & 1 & 0 & 0 & 10 \\ 3 & 4 & 0 & 1 & 0 & 6 \\ \hline -10 & -1 & 0 & 0 & 1 & 0 \end{array}\right]. \\ \uparrow \end{array}$$

Note that this tableau does not correspond to a maximum, since the left part of the bottom row has negative entries. So we pivot to create a new tableau. Since -10 is the most negative entry in the last row, we choose the first column as the pivot column. To determine the pivot element, we compute ratios:

$$\begin{array}{c} \\ u \\ v \\ M \end{array} \begin{array}{c} \begin{array}{ccccc|c} x & y & u & v & M & \end{array} \\ \left[\begin{array}{ccccc|c} 1 & 2 & 1 & 0 & 0 & 10 \\ \textcircled{3} & 4 & 0 & 1 & 0 & 6 \\ \hline -10 & -1 & 0 & 0 & 1 & 0 \end{array}\right] \\ \uparrow \end{array} \begin{array}{c} \text{Ratios} \\ 10/1 = 10 \\ 6/3 = 2 \\ \\ \end{array}$$

The smallest ratio is 2, so we pivot about 3, which we have circled. The new tableau is therefore

$$\begin{array}{c} \\ u \\ x \\ M \end{array} \begin{array}{c} \begin{array}{ccccc|c} x & y & u & v & M & \end{array} \\ \left[\begin{array}{ccccc|c} 0 & \frac{1}{3} & 1 & -\frac{1}{3} & 0 & 8 \\ 1 & \frac{4}{3} & 0 & \frac{1}{3} & 0 & 2 \\ \hline 0 & \frac{37}{3} & 0 & \frac{10}{3} & 1 & 20 \end{array}\right]. \end{array}$$

Note that this tableau corresponds to a maximum, since there are no negative entries in the left part of the last row. The solution corresponding to the tableau is

$$x = 2, \quad y = 0, \quad u = 8, \quad v = 0, \quad M = 20.$$

Therefore, the objective function assumes its maximum value of 20 when $x = 2$ and $y = 0$.

Let us now check that the simplex method yields the same result as the graphical method of Chapter 2 by reworking a problem from there.

EXAMPLE 2 Use the simplex method to solve the furniture manufacturing problem of Section 1 of Chapter 2

Solution Here x represents the number of chairs and y the number of sofas to be produced each day. The daily profit is $80x + 70y$ dollars, and the limitations imposed by available labor are expressed by the constraints

$$\begin{cases} 6x + 3y \le 96 \\ x + y \le 18 \\ 2x + 6y \le 72 \\ x \ge 0, \quad y \ge 0. \end{cases}$$

The linear system is then

$$\begin{cases} 6x + 3y + u & = 96 \\ x + y + v & = 18 \\ 2x + 6y + w & = 72 \\ -80x - 70y + M & = 0. \end{cases}$$

The simplex method then proceeds as follows:

$$\begin{array}{c} \\ u \\ v \\ w \\ M \\ \\ \end{array}
\begin{array}{cccccc|c} x & y & u & v & w & M & \\ \hline \text{\textcircled{6}} & 3 & 1 & 0 & 0 & 0 & 96 \\ 1 & 1 & 0 & 1 & 0 & 0 & 18 \\ 2 & 6 & 0 & 0 & 1 & 0 & 72 \\ \hline -80 & -70 & 0 & 0 & 0 & 1 & 0 \\ \uparrow & & & & & & \end{array}
\quad \begin{array}{l} \\ 96/6 = 16 \\ 18/1 = 18 \\ 72/2 = 36 \\ \\ \\ \end{array}$$

$$\begin{array}{c} \\ x \\ v \\ w \\ M \\ \\ \end{array}
\begin{array}{cccccc|c} x & y & u & v & w & M & \\ \hline 1 & \frac{1}{2} & \frac{1}{6} & 0 & 0 & 0 & 16 \\ 0 & \text{\textcircled{$\frac{1}{2}$}} & -\frac{1}{6} & 1 & 0 & 0 & 2 \\ 0 & 5 & -\frac{1}{3} & 0 & 1 & 0 & 40 \\ \hline 0 & -30 & \frac{40}{3} & 0 & 0 & 1 & 1280 \\ & \uparrow & & & & & \end{array}
\quad \begin{array}{l} \\ 16/\frac{1}{2} = 32 \\ 2/\frac{1}{2} = 4 \\ 40/5 = 8 \\ \\ \\ \end{array}$$

$$\begin{array}{c} \\ x \\ y \\ w \\ M \end{array}
\begin{array}{cccccc|c} x & y & u & v & w & M & \\ \hline 1 & 0 & \frac{1}{3} & -1 & 0 & 0 & 14 \\ 0 & 1 & -\frac{1}{3} & 2 & 0 & 0 & 4 \\ 0 & 0 & \frac{4}{3} & -10 & 1 & 0 & 20 \\ \hline 0 & 0 & \frac{10}{3} & 60 & 0 & 1 & 1400 \end{array}$$

This last tableau corresponds to maximum profit, and the solution is

$$x = 14, \qquad y = 4, \qquad u = 0, \qquad v = 0, \qquad w = 20, \qquad M = 1400.$$

Thus the maximum profit of \$1400 occurs when $x = 14$ and $y = 4$.

The simplex method can be used to solve problems in any number of variables. Let us illustrate the method for three variables.

EXAMPLE 3 Maximize the objective function $x + 2y + z$ subject to the constraints

$$\begin{cases} x - y + 2z \leq 10 \\ 2x + y + 3z \leq 12 \\ x \geq 0, \quad y \geq 0, \quad z \geq 0. \end{cases}$$

Solution We determined the corresponding linear system in Example 1 of Section 1:

$$\begin{cases} x - y + 2z + u = 10 \\ 2x + y + 3z + v = 12 \\ -x - 2y - z + M = 0. \end{cases}$$

So the simplex method works as follows:

$$\begin{array}{c} \\ u \\ v \\ M \end{array} \begin{array}{c} \begin{array}{cccccc|c} x & y & z & u & v & M & \\ \end{array} \\ \left[\begin{array}{cccccc|c} 1 & -1 & 2 & 1 & 0 & 0 & 10 \\ 2 & \textcircled{1} & 3 & 0 & 1 & 0 & 12 \\ \hline -1 & -2 & -1 & 0 & 0 & 1 & 0 \end{array}\right] \\ \uparrow \end{array} \quad 12/1 = 12 \quad \text{(smallest ratio)}$$

$$\begin{array}{c} \\ u \\ y \\ M \end{array} \begin{array}{c} \begin{array}{cccccc|c} x & y & z & u & v & M & \\ \end{array} \\ \left[\begin{array}{cccccc|c} 3 & 0 & 5 & 1 & 1 & 0 & 22 \\ 2 & 1 & 3 & 0 & 1 & 0 & 12 \\ \hline 3 & 0 & 5 & 0 & 2 & 1 & 24 \end{array}\right] \end{array}.$$

Thus the solution of the original problem is: $x = 0$, $y = 12$, $z = 0$ yields the maximum value of the objective function $x + 2y + z$. The maximum value is 24.

PRACTICE PROBLEMS 2

1. Which of these simplex tableaux has a solution which corresponds to a maximum for the associated linear programming problem?

(a)
$$\begin{array}{c} \begin{array}{ccccc|c} x & y & u & v & M & \\ \end{array} \\ \left[\begin{array}{ccccc|c} 3 & 1 & 0 & 1 & 0 & 5 \\ 2 & 0 & 0 & 0 & 1 & 0 \\ \hline -1 & -2 & 1 & 0 & 0 & 3 \end{array}\right] \end{array}$$

(b)
$$\begin{array}{c} \begin{array}{ccccc|c} x & y & u & v & M & \\ \end{array} \\ \left[\begin{array}{ccccc|c} 2 & 1 & 0 & 11 & 0 & 10 \\ 1 & 0 & 1 & 7 & 0 & 1 \\ \hline 1 & 0 & 0 & 4 & 1 & -2 \end{array}\right] \end{array}$$

2. Suppose that in the solution of a linear programming problem by the simplex method we encounter the following simplex tableau. What is the next step in the solution?

$$\begin{array}{c} \begin{array}{ccccc|c} x & y & u & v & M & \end{array} \\ \left[\begin{array}{ccccc|c} 0 & 4 & 1 & 2 & 0 & 4 \\ 1 & 5 & 0 & 1 & 0 & 9 \\ \hline 0 & 2 & 0 & -3 & 1 & 6 \end{array}\right] \end{array}$$

EXERCISES 2

For each of the following simplex tableaux:

(a) Compute the next pivot element.

(b) Determine the next tableau.

(c) Determine the particular solution corresponding to the tableau of part (b).

1.
$$\begin{array}{c} \begin{array}{ccccc|c} x & y & u & v & M & \end{array} \\ \left[\begin{array}{ccccc|c} 6 & 2 & 1 & 0 & 0 & 10 \\ 1 & 3 & 0 & 1 & 0 & 6 \\ \hline -4 & -12 & 0 & 0 & 1 & 0 \end{array}\right] \end{array}$$

2.
$$\begin{array}{c} \begin{array}{ccccc|c} x & y & u & v & M & \end{array} \\ \left[\begin{array}{ccccc|c} 1 & 0 & 3 & 1 & 0 & 5 \\ 0 & 1 & 2 & 0 & 0 & 12 \\ \hline -6 & 0 & 5 & 0 & 1 & 10 \end{array}\right] \end{array}$$

3.
$$\begin{array}{c} \begin{array}{ccccc|c} x & y & u & v & M & \end{array} \\ \left[\begin{array}{ccccc|c} 5 & 12 & 1 & 0 & 0 & 12 \\ 15 & 10 & 0 & 1 & 0 & 5 \\ \hline 4 & -2 & 0 & 0 & 1 & 0 \end{array}\right] \end{array}$$

4.
$$\begin{array}{c} \begin{array}{ccccc|c} x & y & u & v & M & \end{array} \\ \left[\begin{array}{ccccc|c} 0 & 6 & 3 & 1 & 0 & 5 \\ 1 & -5 & 2 & 0 & 0 & 8 \\ \hline 0 & 20 & -10 & 0 & 1 & 22 \end{array}\right] \end{array}$$

Solve the following linear programming problems using the simplex method.

5. Maximize $x + 3y$ subject to the constraints

$$\begin{cases} x + y \le 7 \\ x + 2y \le 10 \\ x \ge 0, \quad y \ge 0. \end{cases}$$

6. Maximize $x + 2y$ subject to the constraints

$$\begin{aligned} -x + y &\le 100 \\ 6x + 6y &\le 1200 \\ x \ge 0, \quad y &\ge 0. \end{aligned}$$

7. Maximize $4x + 2y$ subject to the constraints

$$\begin{cases} 5x + y \le 80 \\ 3x + 2y \le 76 \\ x \ge 0, \quad y \ge 0. \end{cases}$$

8. Maximize $2x + 6y$ subject to the constraints

$$\begin{cases} -x + 8y \le 160 \\ 3x - y \le 3 \\ x \ge 0, \quad y \ge 0. \end{cases}$$

9. Maximize $x + 3y + 5z$ subject to the constraints

$$\begin{cases} x + 2z \le 10 \\ 3y + z \le 24 \\ x \ge 0, \quad y \ge 0, \quad z \ge 0. \end{cases}$$

10. Maximize $-x + 8y + z$ subject to the constraints

$$\begin{cases} x - 2y + 9z \le 10 \\ y + 4z \le 12 \\ x \ge 0, \quad y \ge 0, \quad z \ge 0. \end{cases}$$

11. Maximize $2x + 3y$ subject to the constraints

$$\begin{cases} 5x + y \le 30 \\ 3x + 2y \le 60 \\ x + y \le 50 \\ x \ge 0, \quad y \ge 0. \end{cases}$$

12. Maximize $10x + 12y + 10z$ subject to the constraints

$$\begin{cases} x - 2y \le 6 \\ 3x + z \le 9 \\ y + 3z \le 12 \\ x \ge 0, \quad y \ge 0, \quad z \ge 0. \end{cases}$$

13. Maximize $6x + 7y + 300$ subject to the constraints

$$\begin{cases} 2x + 3y \le 400 \\ x + y \le 150 \\ x \ge 0, \quad y \ge 0. \end{cases}$$

14. Maximize $10x + 20y + 50$ subject to the constraints

$$\begin{cases} x + y \le 10 \\ 5x + 2y \le 20 \\ x \ge 0, \quad y \ge 0. \end{cases}$$

15. Suppose that a furniture manufacturer makes chairs, sofas, and tables. The amounts of labor of various types as well as the relative availability of each type is summarized by the following chart:

	Chair	*Sofa*	*Table*	*Daily labor available (man-hours)*
Carpentry	6	3	8	768
Finishing	1	1	2	144
Upholstery	2	5	0	216

The profit per chair is \$80, per sofa \$70, and per table \$120. How many pieces of each type of furniture should be manufactured each day in order to maximize the profit?

16. A stereo store sells three brands of stereo system, brands A, B, and C. It can sell a total of 100 stereo systems per month. Brands A, B, and C take up, respectively, 5, 4, and 4 cubic feet of warehouse space and a maximum of 480 cubic feet of warehouse space is available. Brands A, B, and C generate sales commissions of \$40, \$20, and \$30, respectively, and \$3200 is available to pay the sales commissions. The profit generated from the sale of each brand is \$70, \$210, and \$140, respectively. How many of each brand of stereo system should be sold to maximize the profit?

17. A furniture manufacturer produces small sofas, large sofas, and chairs. The profits per item are, respectively, \$60, \$60, and \$50. The pieces of furniture require the following numbers of man-hours for their manufacture:

	Carpentry	*Upholstery*	*Finishing*
Small sofas	10	30	20
Large sofas	10	30	0
Chairs	10	10	10

The following amounts of labor are available per month: carpentry at most 1200 hours, upholstery at most 3000 hours, and finishing at most 1800 hours. How many each of small sofas, large sofas, and chairs should be manufactured to maximize the profit?

18. Maximize $60x + 90y + 300z$ subject to the constraints

$$\begin{cases} x + y + z \leq 600 \\ x + 3y \leq 600 \\ 2x + z \leq 900 \\ x \geq 0, \quad y \geq 0, \quad z \geq 0. \end{cases}$$

19. Maximize $200x + 500y$ subject to the constraints

$$\begin{cases} x + 4y \leq 300 \\ x + 2y \leq 200. \end{cases}$$

SOLUTIONS TO PRACTICE PROBLEMS 2

1. (a) Does not correspond to a maximum, since among the entries $-1, -2, 1, 0, 0$ on the last row, at least one is negative.

   (b) Corresponds to a maximum since none of the entries $1, 0, 0, 4, 1$ of the last row is negative. Note that it does not matter that the entry -2 in the right-hand corner of the matrix is negative. This number gives the value of M. In this example -2 is as large as M can become.

2. First choose the column corresponding to the most negative entry of the final row, that is, the fourth column. For each entry in the fourth column which is above the horizontal line, compute the ratio with the sixth column. The smallest ratio is 2 and appears in the first row, so the next operation is to pivot around the 2 in the first row of the fourth column.

$$\left[\begin{array}{ccccc|c} 0 & 4 & 1 & \textcircled{2} & 0 & 4 \\ 1 & 5 & 0 & 1 & 0 & 9 \\ \hline 0 & 2 & 0 & -3 & 1 & 6 \end{array}\right] \begin{array}{l} 4/2 = 2 \\ 9/1 = 9 \\ \\ \end{array}$$
$$\uparrow$$

3.3 The Simplex Method, II—Minimum Problems

In the preceding section we developed the simplex method and applied it to a number of problems. However, throughout we restricted ourselves to linear programming problems in standard form. Recall that such problems satisfied three properties: (1) the objective function is to be maximized; (2) each variable must be ≥ 0; and (3) all constraints other than those implied by (2) must be of the form

$$[\text{linear polynomial}] \leq [\text{nonnegative constant}].$$

In this section we shall do what we can to relax these restrictions.

Let us begin with restriction (3). This could be violated in two ways. First, the constant on the right-hand side of one or more constraints could be negative. Thus, for example, one constraint might be

$$x - y \leq -2.$$

A second way in which restriction (3) can be violated is for some constraints to involve $\geq$ rather than $\leq$. An example of such a constraint is

$$2x + 3y \geq 5.$$

However, we can convert such a constraint into one involving $\leq$ by multiplying both sides of the inequality by -1:

$$-2x - 3y \leq -5.$$

Of course, the right-hand constant is no longer nonnegative. Thus, if we allow negative constants on the right, we can write all constraints in the form

$$[\text{linear polynomial}] \leq [\text{constant}].$$

Henceforth, the first step in solving a linear programming problem will be to write the constraints in this form. Let us now see how to deal with the phenomenon of negative constants.

EXAMPLE 1 Maximize the objective function $5x + 10y$ subject to the constraints

$$\begin{cases} x + y \leq 20 \\ 2x - y \geq 10 \\ x \geq 0, \quad y \geq 0. \end{cases}$$

Solution The first step is to put the second constraint into $\leq$ form. Multiply the second inequality by -1 to obtain

$$\begin{cases} x + y \leq 20 \\ -2x + y \leq -10 \\ x \geq 0, \quad y \geq 0. \end{cases}$$

Just as before, write this as a linear system:

$$\begin{cases} x + y + u = 20 \\ -2x + y + v = -10 \\ -5x - 10y + M = 0. \end{cases}$$

From the linear system construct the simplex tableau:

$$\begin{array}{c} \\ u \\ v \\ M \end{array} \begin{array}{c} \begin{array}{ccccc|c} x & y & u & v & M & \end{array} \\ \left[\begin{array}{ccccc|c} 1 & 1 & 1 & 0 & 0 & 20 \\ -2 & 1 & 0 & 1 & 0 & -10 \\ \hline -5 & -10 & 0 & 0 & 1 & 0 \end{array}\right] \end{array}.$$

Everything would proceed exactly as before, except that the right-hand column has a -10 in it. This means that the initial value for v is -10, which violates the condition that all variables be ≥ 0. Before we can apply the simplex method of Section 2, we must first put the tableau into standard form. This can be done by pivoting so as to remove the negative entry in the right column.

We choose the pivot element as follows. Look along the left side of the -10 row of the tableau and locate any negative entry. There is only one: -2. Use the column containing the -2—column 1—as the pivot column. Now compute ratios as before:*

$$\begin{array}{c} \\ u \\ v \\ M \\ \\ \end{array} \begin{array}{c} \begin{array}{ccccc|c} x & y & u & v & M & \end{array} \\ \left[\begin{array}{ccccc|c} 1 & 1 & 1 & 0 & 0 & 20 \\ \textcircled{-2} & 1 & 0 & 1 & 0 & -10 \\ \hline -5 & -10 & 0 & 0 & 1 & 0 \end{array}\right] \\ \uparrow \qquad\qquad\qquad\qquad\qquad \end{array} \begin{array}{c} \\ 20/1 = 20 \\ -10/-2 = 5 \\ \\ \\ \end{array}$$

* Note, however, that in this circumstance we compute ratios corresponding to both positive *and* negative entries (except the last) in the pivot column, considering further only those ratios which are positive.

The smallest positive ratio is 5, so we choose -2 as the pivot element. The new tableau is

$$\begin{array}{c} \\ u \\ x \\ M \end{array} \begin{array}{c} \begin{array}{ccccc} x & y & u & v & M \end{array} \\ \left[\begin{array}{ccccc|c} 0 & \frac{3}{2} & 1 & \frac{1}{2} & 0 & 15 \\ 1 & -\frac{1}{2} & 0 & -\frac{1}{2} & 0 & 5 \\ \hline 0 & -\frac{25}{2} & 0 & -\frac{5}{2} & 1 & 25 \end{array}\right]. \end{array}$$

Note that all entries in the right-hand column are now nonnegative;* that is, the corresponding solution has all variables ≥ 0. From here on we follow the simplex method for tableaux in standard form:

$$\begin{array}{c} \\ u \\ x \\ M \end{array} \begin{array}{c} \begin{array}{ccccc} x & y & u & v & M \end{array} \\ \left[\begin{array}{ccccc|c} 0 & \textcircled{\tfrac{3}{2}} & 1 & \frac{1}{2} & 0 & 15 \\ 1 & -\frac{1}{2} & 0 & -\frac{1}{2} & 0 & 5 \\ \hline 0 & -\frac{25}{2} & 0 & -\frac{5}{2} & 1 & 25 \end{array}\right] \\ \uparrow \end{array} \quad 15/\tfrac{3}{2} = 10$$

$$\begin{array}{c} \\ y \\ x \\ M \end{array} \begin{array}{c} \begin{array}{ccccc} x & y & u & v & M \end{array} \\ \left[\begin{array}{ccccc|c} 0 & 1 & \frac{2}{3} & \frac{1}{3} & 0 & 10 \\ 1 & 0 & \frac{1}{3} & -\frac{1}{3} & 0 & 10 \\ \hline 0 & 0 & \frac{25}{3} & \frac{5}{3} & 1 & 150 \end{array}\right]. \end{array}$$

So the maximum value of M is 150, which is attained for $x = 10$, $y = 10$.

In summary:

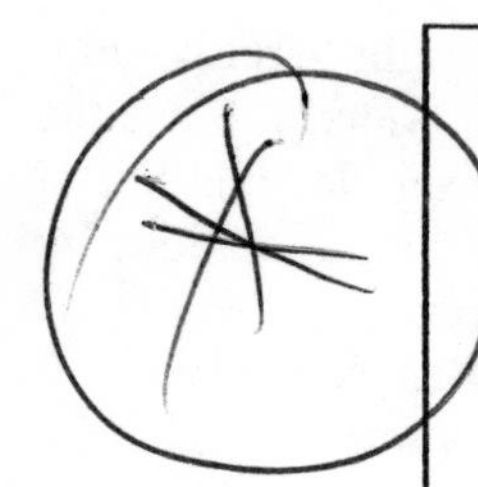

The Simplex Method for Problems in Nonstandard Form

1. If necessary, convert all inequalities (except $x \geq 0$, $y \geq 0$), into the form

$$[\text{linear polynomial}] \leq [\text{constant}].$$

2. If a negative number appears in the upper part of the last column of the simplex tableau, remove it by pivoting.
   (a) Select one of the negative entries in its row. The column containing the entry will be the pivot column.
   (b) Select the pivot element by determining the least of the positive ratios associated to entries in the pivot column (except the bottom entry).
   (c) Pivot.
3. Repeat step 2 until there are no negative entries in the upper part of the right-hand column of the simplex tableau.
4. Proceed to apply the simplex method for tableaux in standard form.

---

* In general, it may be necessary to pivot several times before all elements in the last column are ≥ 0.

The method we have just developed can be used to solve ***minimum*** **problems** as well as maximum problems. Minimizing the objective function f is the same as maximizing $(-1) \cdot f$. This is so since multiplying an inequality by -1 reverses the direction of the inequality sign. Thus, in order to apply our method to a minimum problem, we merely multiply the objective function by -1 and turn the problem into a maximum problem.

EXAMPLE 2 Minimize the objective function $3x + 2y$ subject to the constraints

$$\begin{cases} x + y \geq 10 \\ x - y \leq 15 \\ x \geq 0, \quad y \geq 0. \end{cases}$$

Solution First transform the problem so that the first two constraints are in $\leq$ form:

$$\begin{cases} -x - y \leq -10 \\ x - y \leq 15 \\ x \geq 0, \quad y \geq 0. \end{cases}$$

Instead of minimizing $3x + 2y$, let us maximize $-3x - 2y$. Let $M = -3x - 2y$. Then our initial simplex tableau reads

$$\begin{array}{c} \\ u \\ v \\ M \end{array} \begin{array}{c} \begin{array}{ccccc|c} x & y & u & v & M & \\ \end{array} \\ \left[\begin{array}{ccccc|c} -1 & \textcircled{-1} & 1 & 0 & 0 & -10 \\ 1 & -1 & 0 & 1 & 0 & 15 \\ \hline 3 & 2 & 0 & 0 & 1 & 0 \end{array}\right] \\ \uparrow \end{array} \quad -10/-1 = 10$$

We first eliminate the -10 in the right-hand column. We have a choice of two negative entries in the -10 row. Let us choose the one in the y column. The ratios are then tabulated as above, and we pivot around the circled element. The new tableau is*

$$\begin{array}{c} \\ y \\ v \\ M \end{array} \begin{array}{c} \begin{array}{ccccc|c} x & y & u & v & M & \\ \end{array} \\ \left[\begin{array}{ccccc|c} 1 & 1 & -1 & 0 & 0 & 10 \\ 2 & 0 & -1 & 1 & 0 & 25 \\ \hline 1 & 0 & 2 & 0 & 1 & -20 \end{array}\right] \end{array}.$$

Since all entries in the bottom row, except the last, are positive, this tableau corresponds to a maximum. Thus the maximum value of $-3x - 2y$ (subject to the constraints) is -20 and this value occurs for $x = 0$, $y = 10$. Thus the *minimum* value of $3x + 2y$ subject to the constraints is 20.

Let us now rework an applied problem previously treated (see Example 2 in Section 3 of **Chapter 2**),this time using the simplex method. For easy reference we restate the problem.

* Note that we do not need the last entry in the last column positive. We require *only* that x, y, u, and v be ≥ 0.

EXAMPLE 3 (*Transportation Problem*) Suppose that a TV dealer has stores in Annapolis and Rockville and warehouses in College Park and Baltimore. The cost of shipping sets from College Park to Annapolis is \$6 per set; from College Park to Rockville, \$3; from Baltimore to Annapolis, \$9; and from Baltimore to Rockville, \$5. Suppose that the Annapolis store orders 25 TV sets and the Rockville store 30. Further suppose that the College Park warehouse has a stock of 45 sets, and the Baltimore warehouse 40. What is the most economical way to supply the requested TV sets to the two stores?

Solution via the Simplex Method As in the previous solution, let x be the number of sets shipped from College Park to Rockville, and y the number shipped from College Park to Annapolis. The flow of sets is depicted in Fig. 1.

Exactly as in our previous solution, we reduce the problem to the following algebraic form: Minimize $375 - 2x - 3y$ subject to the constraints

$$\begin{cases} x \leq 30, \quad y \leq 25 \\ x + y \geq 15 \\ x + y \leq 45 \\ x \geq 0, \quad y \geq 0. \end{cases}$$

Two changes are needed. First, instead of minimizing $375 - 2x - 2y$ we maximize $-(375 - 2x - 3y) = 2x + 3y - 375$. Second, we write the constraint $x + y \geq 15$ in the form

$$-x - y \leq -15.$$

With these changes made, we can write down the linear system:

$$\begin{cases} x \quad\quad + t \quad\quad\quad\quad\quad\quad = 30 \\ \quad\quad y \quad + u \quad\quad\quad\quad = 25 \\ -x - y \quad\quad\quad + v \quad\quad\quad = -15 \\ x + y \quad\quad\quad\quad\quad + w \quad = 45 \\ -2x - 3y \quad\quad\quad\quad\quad\quad + M = -375. \end{cases}$$

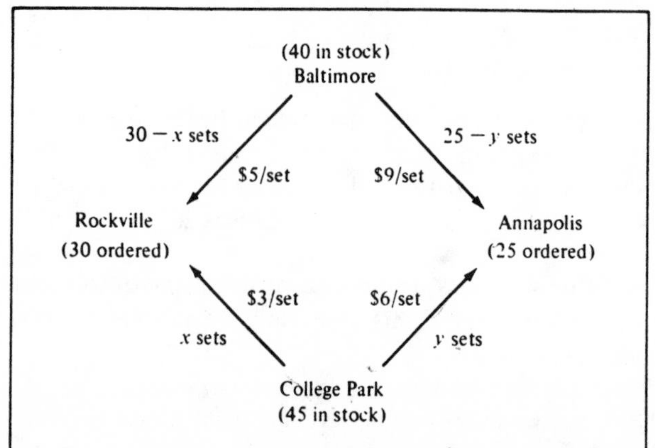

FIGURE 1

From here on we follow our routine procedure in a mechanical way:

$$
\begin{array}{c|ccccccc|c|c}
 & x & y & t & u & v & w & M & & \\
\hline
t & 1 & 0 & 1 & 0 & 0 & 0 & 0 & 30 & 30/1 = 30 \\
u & 0 & 1 & 0 & 1 & 0 & 0 & 0 & 25 & \\
v & \textcircled{-1} & -1 & 0 & 0 & 1 & 0 & 0 & -15 & -15/-1 = 15 \\
w & 1 & 1 & 0 & 0 & 0 & 1 & 0 & 45 & 45/1 = 45 \\
\hline
M & -2 & -3 & 0 & 0 & 0 & 0 & 1 & -375 & \\
 & \uparrow & & & & & & & &
\end{array}
$$

$$
\begin{array}{c|ccccccc|c|c}
 & x & y & t & u & v & w & M & & \\
\hline
t & 0 & -1 & 1 & 0 & \textcircled{1} & 0 & 0 & 15 & 15/1 = 15 \\
u & 0 & 1 & 0 & 1 & 0 & 0 & 0 & 25 & \\
x & 1 & 1 & 0 & 0 & -1 & 0 & 0 & 15 & \\
w & 0 & 0 & 0 & 0 & 1 & 1 & 0 & 30 & 30/1 = 30 \\
\hline
M & 0 & -1 & 0 & 0 & -2 & 0 & 1 & -345 & \\
 & & & & & \uparrow & & & &
\end{array}
$$

$$
\begin{array}{c|ccccccc|c|c}
 & x & y & t & u & v & w & M & & \\
\hline
v & 0 & -1 & 1 & 0 & 1 & 0 & 0 & 15 & \\
u & 0 & 1 & 0 & 1 & 0 & 0 & 0 & 25 & 25/1 = 25 \\
x & 1 & 0 & 1 & 0 & 0 & 0 & 0 & 30 & \\
w & 0 & \textcircled{1} & -1 & 0 & 0 & 1 & 0 & 15 & 15/1 = 15 \\
\hline
M & 0 & -3 & 2 & 0 & 0 & 0 & 1 & -315 & \\
 & & \uparrow & & & & & & &
\end{array}
$$

$$
\begin{array}{c|ccccccc|c|c}
 & x & y & t & u & v & w & M & & \\
\hline
v & 0 & 0 & 0 & 0 & 1 & 1 & 0 & 30 & \\
u & 0 & 0 & \textcircled{1} & 1 & 0 & -1 & 0 & 10 & 10/1 = 10 \\
x & 1 & 0 & 1 & 0 & 0 & 0 & 0 & 30 & 30/1 = 30 \\
y & 0 & 1 & -1 & 0 & 0 & 1 & 0 & 15 & \\
\hline
M & 0 & 0 & -1 & 0 & 0 & 3 & 1 & -270 & \\
 & & & \uparrow & & & & & &
\end{array}
$$

$$
\begin{array}{c|ccccccc|c}
 & x & y & t & u & v & w & M & \\
\hline
v & 0 & 0 & 0 & 0 & 1 & 1 & 0 & 30 \\
t & 0 & 0 & 1 & 1 & 0 & -1 & 0 & 10 \\
x & 1 & 0 & 0 & -1 & 0 & 1 & 0 & 20 \\
y & 0 & 1 & 0 & 1 & 0 & 0 & 0 & 25 \\
\hline
M & 0 & 0 & 0 & 1 & 0 & 2 & 1 & -260
\end{array}
$$

This last tableau corresponds to a maximum. So $2x + 3y - 375$ has a maximum value -260, and therefore $375 - 2x - 3y$ has a minimum value 260. This value occurs when $x = 20$ and $y = 25$. This is in agreement with our previous graphical solution of the problem.

The calculations used in the preceding example are not that much simpler than those in the original solution. Why, then, should we concern ourselves with the simplex method? For one thing, the simplex method is so mechanical in its execution that it is much easier to program for a computer. For another, our previous method was restricted to problems in two variables. However, suppose that the two warehouses were to deliver their TV sets to three or four or perhaps even 100 stores? Our previous method could not be applied. However, the simplex method, although yielding very large matrices and very tedious calculations, is applicable. Indeed, this is the method used by many industrial concerns to optimize distribution of their products.

Some Further Comments on the Simplex Method Our discussion has omitted some of the technical complications arising in the simplex method. A complete discussion of these is beyond the scope of this book. However, let us mention two. First of all, it is possible that a given linear programming problem has more than one solution. This can occur, for example, if there are ties for the choice of pivot column. For instance, if the bottom row of the simplex tableau is

$$\begin{bmatrix} -3 & -7 & 4 & -7 & 1 & 3 \end{bmatrix}, \\ \qquad\;\; \uparrow \qquad\quad \uparrow$$

then -7 is the most negative entry and we may choose as pivot column either the second or fourth. In such a circumstance the pivot column may be chosen arbitrarily. Different choices, however, may lead to different solutions of the problem.

A second difficulty is that a given linear programming problem may have no solution at all. In this case the method will break down at some point. For example, among the ratios at a given stage there may be no nonnegative ones to consider. Then we cannot choose a pivot element. Such a breakdown of the method indicates that the associated linear programming problem has no solution.

PRACTICE PROBLEMS 3

1. Convert the following minimum problem into a maximum problem in standard form: Minimize $3x + 4y$ subject to the constraints

$$\begin{cases} x - y \geq 0 \\ 3x - 4y \geq 0 \\ x \geq 0, \quad y \geq 0. \end{cases}$$

2. Suppose that the solution of a minimum problem yields the final simplex tableau

$$\begin{array}{c} \begin{matrix} x & y & u & v & M & \end{matrix} \\ \left[\begin{array}{rrrrr|r} 1 & 6 & -1 & 0 & 0 & 11 \\ 0 & 5 & 3 & 1 & 0 & 16 \\ \hline 0 & 2 & 4 & 0 & 1 & -40 \end{array}\right]. \end{array}$$

What is the minimum value sought in the original problem?

EXERCISES 3

Solve the following linear programming problems by the simplex method.

1. Maximize $40x + 30y$ subject to the constraints

$$\begin{cases} x + y \le 5 \\ -2x + 3y \ge 12 \\ x \ge 0, \quad y \ge 0. \end{cases}$$

2. Maximize $3x - y$ subject to the constraints

$$\begin{cases} 2x + 5y \le 100 \\ x \ge 10 \\ y \ge 0. \end{cases}$$

3. Minimize $3x + y$ subject to the constraints

$$\begin{cases} x + y \ge 3 \\ 2x \ge 5 \\ x \ge 0, \quad y \ge 0. \end{cases}$$

4. Minimize $3x + 5y + z$ subject to the constraints

$$\begin{cases} x + y + z \ge 20 \\ y + 2z \ge 10 \\ x \ge 0, \quad y \ge 0, \quad z \ge 0. \end{cases}$$

5. Minimize $13x + 4y$ subject to the constraints

$$\begin{cases} y \ge -2x + 11 \\ y \le -x + 10 \\ y \le -\frac{1}{3}x + 6 \\ y \ge -\frac{1}{4}x + 4 \\ x \ge 0, \quad y \ge 0. \end{cases}$$

6. Minimize $500 - 10x - 3y$ subject to the constraints

$$\begin{cases} x + y \le 20 \\ 3x + 2y \ge 50 \\ x \ge 0, \quad y \ge 0. \end{cases}$$

7. A dietician is designing a daily diet that is to contain at least 60 units of protein, 40 units of carbohydrate, and 120 units of fat. The diet is to consist of two types of foods. One serving of food A contains 30 units of protein, 10 units of carbohydrate, and 20 units of fat and costs $3. One serving of food B contains 10 units of protein, 10 units of carbohydrate, and 60 units of fat and costs $1.50. Design the diet that provides the daily requirements at the least cost.

8. A manufacturing company has two plants, each capable of producing radios, television sets, and stereo systems. The daily production capacities of each plant are as follows:

	Plant I	*Plant II*
Radios	10	20
Television sets	30	20
Stereo systems	20	10

Plant I costs \$1500 per day to operate, whereas plant II costs \$1200. How many days should each plant be operated to fill an order for 1000 radios, 1800 television sets, and 1000 stereo systems at the minimum cost?

9. An appliance store sells three brands of color television sets, brands A, B, and C. The profit per set is \$30 for brand A, \$50 for brand B, and \$60 for brand C. The total warehouse space allocated to all brands is sufficient for 600 sets and the inventory is delivered only once per month. At least 100 customers per month will demand brand A, at least 50 will demand brand B, and at least 200 will demand either brand B or brand C. How can the appliance store satisfy all these demands and earn maximum profit?

10. A citizen decides to campaign for the election of a candidate for city council. Her goal is to generate at least 210 votes by a combination of door-to-door canvassing, letter writing, and phone calls. She figures that each hour of door-to-door canvassing will generate four votes, each hour of letter writing will generate two votes, and each hour on the phone will generate three votes. She would like to devote at least seven hours to phone calls and spend at most half of her time at door-to-door canvassing. How much time should she allocate to each task in order to achieve her goal in the least amount of time?

SOLUTIONS TO PRACTICE PROBLEMS 3

1. To minimize $3x + 4y$, we maximize $-(3x + 4y) = -3x - 4y$. So the associated maximum problem is: Maximize $-3x - 4y$ subject to the constraints

$$\begin{cases} -x + y \leq 0 \\ -3x + 4y \leq 0 \\ x \geq 0, \quad y \geq 0. \end{cases}$$

2. The value -40 in the lower right corner gives the solution of the associated *maximum* problem. The minimum value originally sought is the negative of the maximum value—that is, $-(-40) = 40$.

3.4 Duality

Each linear programming problem may be converted into a related linear programming problem called its *dual*. The dual problem is sometimes easier to solve than the original problem and moreover has the same optimum value. Furthermore, the solution of the dual problem often can provide valuable insights into the original problem. In order to understand the relationship between a linear programming problem and its dual, it is best to begin with a concrete example.

PROBLEM A Maximize the objective function $6x + 5y$ subject to the constraints

$$\begin{cases} 4x + 8y \leq 32 \\ 3x + 2y \leq 12 \\ x \geq 0, \quad y \geq 0. \end{cases}$$

The dual of Problem A is the following problem.

PROBLEM B Minimize the objective function $32u + 12v$ subject to the constraints

$$\begin{cases} 4u + 3v \geq 6 \\ 8u + 2v \geq 5 \\ u \geq 0, \quad v \geq 0. \end{cases}$$

The relationship between the two problems is easiest to see if we first display the data from Problem A in a matrix, allowing each nontrivial constraint (i.e., other than $x \geq 0$, $y \geq 0$) to occupy one row. The objective function is written in the last row.

$$\begin{matrix} x & y & \\ \end{matrix}$$
$$\begin{bmatrix} 4 & 8 & 32 \\ 3 & 2 & 12 \\ 6 & 5 & 0 \end{bmatrix}$$

The dual problem is obtained from the columns of the matrix. The final column corresponds to the new objective function and the remaining columns give the constraints. Note that the $\leq$ signs of Problem A become $\geq$ signs in the dual problem.

In a similar fashion, we may form the dual of any linear programming problem using the following procedure.

The Dual Problem

1. **If the given problem is a maximum problem, write all nontrivial constraints using only $\leq$. If the given problem is a minimum problem, write all constraints using only $\geq$. (If an inequality points in the wrong direction, we need only multiply it by -1.)**
2. **Display the data of the given problem in matrix form, with each nontrivial constraint occupying one row and the objective function occupying the final row.**
3. **The objective function of the dual problem is formed from the final column of the matrix. If the original problem involved a maximum, then the dual problem involves a minimum; if the original problem involved a minimum, then the dual problem involves a maximum.**
4. **The nontrivial constraints of the dual problem are formed from the remaining columns of the matrix. The inequality signs are the reverse of those in the original problem. In addition, the variables are constrained to be nonnegative.**

EXAMPLE 1 Determine the dual of the following linear programming problem. Minimize $18x + 20y + 2z$ subject to the constraints

$$\begin{cases} 3x - 5y - 2z \leq 4 \\ 6x \qquad\;\; - 8z \geq 9 \\ x \geq 0, \quad y \geq 0, \quad z \geq 0. \end{cases}$$

Solution Since the given problem is a minimization, all inequalities must be written using the inequality sign $\geq$. To put the first inequality in this form, we must multiply by -1 to obtain

$$-3x + 5y + 2z \geq -4.$$

We now display the data in a matrix.

$$\begin{array}{c} \begin{array}{ccc} x & y & z \end{array} \\ \begin{bmatrix} -3 & 5 & 2 & -4 \\ 6 & 0 & -8 & 9 \\ 18 & 20 & 2 & 0 \end{bmatrix}. \end{array}$$

We form the dual problem from the columns of this matrix. The objective function is obtained from the last column: $-4u + 9v$. And since the given problem is a minimization, the objective function of the dual problem is to be maximized. The constraints are obtained from the remaining columns:

$$\begin{cases} -3u + 6v \leq 18 \\ 5u \qquad\quad \leq 20 \\ 2u - 8v \leq 2 \\ u \geq 0, \quad v \geq 0. \end{cases}$$

Let us now return to Problems A and B in order to examine the connection between the solutions of a linear programming problem and its dual problem. Problems A and B both involve two variables and hence can be solved by the geometric method of Chapter 2. Figure 1 shows their respective feasible sets and the vertices that yield the optimum values of the objective functions. The feasible sets do not look alike and the optimal vertices are different. However, both problems have the same optimum value, 27. The relationship between the two problems is brought into even sharper focus by looking at the final tableaux that

FIGURE 1

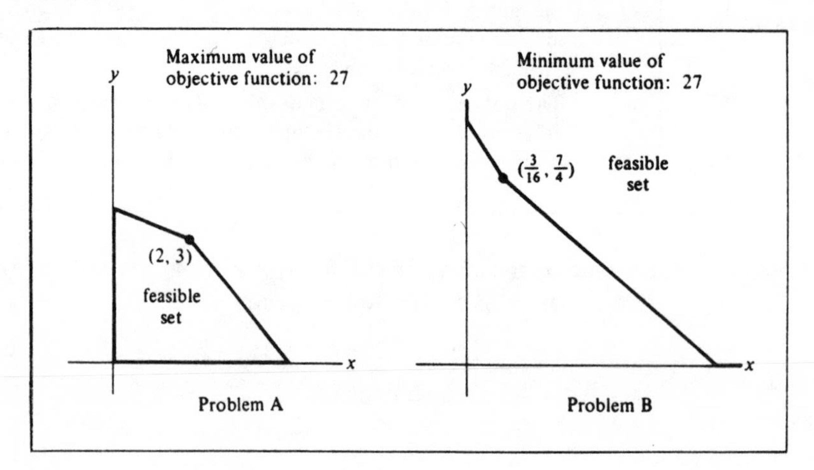

arise when the two problems are solved by the simplex method. (*Note:* In Problem B, the original variables are u and v and the slack variables have been named x and y.)

Final Tableaux

	x	y	u	v	M	
y	0	1	$\frac{3}{16}$	$-\frac{1}{4}$	0	3
x	1	0	$-\frac{1}{8}$	$\frac{1}{2}$	0	2
	0	0	$\frac{3}{16}$	$\frac{7}{4}$	1	27

Problem A

	u	v	x	y	M	
v	0	1	$-\frac{1}{2}$	$\frac{1}{4}$	0	$\frac{7}{4}$
u	1	0	$\frac{1}{8}$	$-\frac{3}{16}$	0	$\frac{3}{16}$
	0	0	2	3	1	-27

Problem B

The final tableau for Problem A contains the solution to Problem B ($u = \frac{3}{16}$, $v = \frac{7}{4}$) in the final entries of the u and v columns. Similarly, the final tableau for Problem B gives the solution to Problem A ($x = 2$, $y = 3$) in the final entries of its x and y columns. This situation always occurs. The solutions to a linear programming problem and its dual problem may be obtained simultaneously by solving just one of the problems using the simplex method and applying the following theorem

> *Fundamental Theorem of Duality* At their respective optimal points, the objective functions of a linear programming problem and its dual problem have the same values. Furthermore, the solution of one of these problems by the simplex method yields the solution of the other as the final entries in the columns for the slack variables.

EXAMPLE 2 Solve the linear programming problem of Example 1 by applying the simplex method to its dual problem.

Solution In the solution to Example 1, the dual problem is: Maximize $-4u + 9v$ subject to the constraints

$$\begin{cases} -3u + 6v \leq 18 \\ 5u \qquad\quad \leq 20 \\ 2u - 8v \leq 2 \\ \quad u \geq 0, \quad v \geq 0. \end{cases}$$

Since there are three nontrivial inequalities, the simplex method calls for three slack variables. Denote the slack variables by x, y, and z. Let $M = -4u + 9v$ and apply the simplex method.

$$
\begin{array}{c}
\\ x \\ y \\ z \\ \\
\end{array}
\begin{array}{cccccc|c}
u & v & x & y & z & M & \\
\hline
-3 & \textcircled{6} & 1 & 0 & 0 & 0 & 18 \\
5 & 0 & 0 & 1 & 0 & 0 & 20 \\
2 & -8 & 0 & 0 & 1 & 0 & 2 \\
\hline
4 & -9 & 0 & 0 & 0 & 1 & 0
\end{array}
$$

$$
\begin{array}{c}
\\ v \\ y \\ z \\ \\
\end{array}
\begin{array}{cccccc|c}
u & v & x & y & z & M & \\
\hline
-\frac{1}{2} & 1 & \frac{1}{6} & 0 & 0 & 0 & 3 \\
\textcircled{5} & 0 & 0 & 1 & 0 & 0 & 20 \\
-2 & 0 & \frac{4}{3} & 0 & 1 & 0 & 26 \\
\hline
-\frac{1}{2} & 0 & \frac{3}{2} & 0 & 0 & 1 & 27
\end{array}
$$

$$
\begin{array}{c}
\\ v \\ u \\ z \\ \\
\end{array}
\begin{array}{cccccc|c}
u & v & x & y & z & M & \\
\hline
0 & 1 & \frac{1}{6} & \frac{1}{10} & 0 & 0 & 5 \\
1 & 0 & 0 & \frac{1}{5} & 0 & 0 & 4 \\
0 & 0 & \frac{4}{3} & \frac{2}{5} & 1 & 0 & 34 \\
\hline
0 & 0 & \frac{3}{2} & \frac{1}{10} & 0 & 1 & 29
\end{array}
$$

Since the maximum value of the dual problem is 29, we know that the minimum value of the original problem is also 29. Looking at the last row of the final tableau, we conclude that this minimum value is assumed when $x = \frac{3}{2}$, $y = \frac{1}{10}$, and $z = 0$.

In Example 2, the dual problem was easier to solve than the original problem given in Example 1. Thus, we see how consideration of the dual problem may simplify the solution of linear programming problems in some cases.

Economic Significance of the Solution of the Dual It may seem that the entire discussion of duality is a mathematical curiosity which is useful for calculation, but has no further "real-world" meaning. But it would be very irresponsible if we left you with this mistaken impression. To illustrate the meaning of the solution of the dual problem, let's reconsider the furniture manufacturing problem of the preceding chapter. Recall that this problem asked us to maximize the profit from the sale of x chairs and y sofas subject to limitations on the amount of labor available for carpentry, upholstery, and finishing. In mathematical terms, the problem required us to maximize $80x + 70y$ subject to the constraints

$$
\begin{cases}
6x + 3y \leq 96 \\
x + y \leq 18 \\
2x + 6y \leq 72 \\
x \geq 0, \quad y \geq 0.
\end{cases}
$$

The furniture manufacturing problem was solved by the simplex method in Section 2. From the last row of the final tableau, we conclude that the optimum value of the dual problem occurs when $u = \frac{10}{3}$, $v = 60$, and $w = 0$. (Here u was the slack variable for the first inequality, v for the second, and w for the third.) The numbers $\frac{10}{3}$, 60, and 0 have the following significance: In the event that one additional man-hour becomes available for carpentry, the profit could be increased by $\frac{10}{3}$ dollars. Similarly, an additional man-hour for upholstery could result in an increase in profit of 60 dollars. However, there can be no increase in profit due to the application of an additional man-hour of finishing.

More precisely, suppose that the number 96 were changed by h man-hours, where h is either positive or negative. If h is not too large,* then the maximum profit will change by $\frac{10}{3}$ times h. For instance, a decrease of $\frac{1}{2}$ man-hour available for carpentry ($h = -\frac{1}{2}$) results in a change in maximum profit of $(-\frac{1}{2})(\frac{10}{3}) = -\frac{5}{3}$ dollars.

An Economic Interpretation of the Dual Problem Here, also, we shall consider the furniture manufacturing problem. Its dual problem is to minimize $96u + 18v + 72w$ subject to the constraints

$$\begin{cases} 6u + v + 2w \geq 80 \\ 3u + v + 6w \geq 70 \\ u \geq 0, \quad v \geq 0, \quad w \geq 0. \end{cases}$$

The variables u, v, and w can be assigned a meaning so that the dual problem has a significant interpretation in terms of the original problem.

First, recall the following table of data (man-hours except as noted):

	Chair	*Sofa*	*Available labor*
Carpentry	6	3	96
Finishing	1	1	18
Upholstery	2	6	72
Profit	\$80	\$70	

Suppose that we have an opportunity to hire out all our workers. Suppose that hiring out the carpenters will yield a profit of u dollars per hour, the finishers v dollars per hour, and the upholsterers w dollars per hour. Of course, u, v, and w must all be ≥ 0. However, there are other constraints which we should reasonably impose. Any scheme for hiring out the workers should generate at least as much profit as is currently being generated in the construction of chairs and sofas. In terms of the potential profits from hiring the workers out, the labor involved in constructing a chair will generate

$$6u + v + 2w$$

dollars of profit. And this amount should be at least equal to the \$80 profit that could be earned by using the labor to construct a chair. That is, we have the

* The number h must be such that the new problem will have the same group II variables in its final tableau as the original problem. Sophisticated computer programs for solving linear programming problems often provide the acceptable range of values for h.

constraint

$$6u + v + 2w \geq 80.$$

Similarly, considering the labor involved in building a sofa, we derive the constraint

$$3u + v + 6w \geq 70.$$

Since there are available 96 hours of carpentry, 18 hours of finishing, and 72 hours of upholstery, the total profit from hiring out the workers would be

$$96u + 18v + 72w.$$

Thus the problem of determining the least acceptable profit from hiring out the workers is equivalent to the following: Minimize $96u + 18v + 72w$ subject to the constraints

$$\begin{cases} 6u + v + 2w \geq 80 \\ 3u + v + 6w \geq 70 \\ u \geq 0, \quad v \geq 0, \quad w \geq 0. \end{cases}$$

This is just the dual of the furniture manufacturing problem. The values u, v, and w are measures of the value of an hour's labor by each type of worker. Economists often refer to them as *shadow profits*. The fundamental theorem of duality asserts that the minimum acceptable profit that can be achieved by hiring the workers out is equal to the maximum profit that can be generated if they make furniture.

Matrices, Linear Programming, and Duality Linear programming problems can be neatly formulated in terms of matrices. Such a formulation is easily transformed into a matrix statement of the dual problem. To introduce the matrix formulation of a linear programming problem, we need the concept of inequality for matrices.

Let A and B be two matrices of the same size. We say that A less than or equal to B (denoted $A \leq B$) if each entry of A is less than or equal to the corresponding entry of B. For instance, we have the following matrix inequalities:

$$\begin{bmatrix} 2 & -3 \\ \frac{1}{2} & 0 \end{bmatrix} \leq \begin{bmatrix} 5 & -1 \\ 1 & 0 \end{bmatrix} \quad \text{and} \quad \begin{bmatrix} 5 \\ 6 \end{bmatrix} \leq \begin{bmatrix} 8 \\ 9 \end{bmatrix}.$$

The symbol $\geq$ has an analogous meaning for matrices.

EXAMPLE 3 Let

$$A = \begin{bmatrix} 6 & 3 \\ 1 & 1 \\ 2 & 6 \end{bmatrix}, \quad B = \begin{bmatrix} 96 \\ 18 \\ 72 \end{bmatrix}, \quad C = [80 \quad 70], \quad X = \begin{bmatrix} x \\ y \end{bmatrix}.$$

Carry out the indicated matrix multiplications in the following statement: Maximize CX subject to the constraints $AX \leq B$, $X \geq \mathbf{0}$. (Here $\mathbf{0}$ is the zero matrix.)

Solution $CX = [80 \quad 70]\begin{bmatrix} x \\ y \end{bmatrix} = [80x + 70y].$

$$AX = \begin{bmatrix} 6 & 3 \\ 1 & 1 \\ 2 & 6 \end{bmatrix}\begin{bmatrix} x \\ y \end{bmatrix} = \begin{bmatrix} 6x + 3y \\ x + y \\ 2x + 6y \end{bmatrix}.$$

$$AX \leq B \text{ means} \begin{bmatrix} 6x + 3y \\ x + y \\ 2x + 6y \end{bmatrix} \leq \begin{bmatrix} 96 \\ 18 \\ 72 \end{bmatrix} \quad \text{or} \quad \begin{cases} 6x + 3y \leq 96 \\ x + y \leq 18 \\ 2x + 6y \leq 72. \end{cases}$$

$$X \geq 0 \text{ means} \begin{bmatrix} x \\ y \end{bmatrix} \geq \begin{bmatrix} 0 \\ 0 \end{bmatrix} \quad \text{or} \quad \begin{cases} x \geq 0 \\ y \geq 0. \end{cases}$$

Hence the statement "Maximize CX subject to the constraints $AX \leq B$, $X \geq \mathbf{0}$" is a matrix formulation of the furniture manufacturing problem.

EXAMPLE 4 Express the dual of the furniture manufacturing problem in matrix form using the matrices of Example 3.

Solution The dual problem is: Minimize $96u + 18v + 72w$ subject to the constraints

$$\begin{cases} 6u + v + 2w \geq 80 \\ 3u + v + 6w \geq 70 \\ u \geq 0, \quad v \geq 0, \quad w \geq 0. \end{cases}$$

Let U be the matrix $[u \quad v \quad w]$ and A, B, C the matrices of Example 3. Then

$$UB = [u \quad v \quad w]\begin{bmatrix} 96 \\ 18 \\ 72 \end{bmatrix} = [96u + 18v + 72w]$$

and

$$UA = [u \quad v \quad w]\begin{bmatrix} 6 & 3 \\ 1 & 1 \\ 2 & 6 \end{bmatrix} = [6u + v + 2w \quad 3u + v + 6w].$$

Hence the dual problem may be stated: Minimize UB subject to the constraints $UA \geq C$, $U \geq \mathbf{0}$.

Example 4 is a particular instance of the following result.

Matrix formulation of the dual Let

$$X = \begin{bmatrix} x \\ y \\ z \\ \vdots \end{bmatrix}$$

and A, B, C be matrices of appropriate sizes. The linear programming problem "Maximize CX subject to the constraints $AX \leq B, X \geq \mathbf{0}$" has as its dual "Minimize UB subject to the constraints $UA \geq C, U \geq \mathbf{0}$, where $U = [u \quad v \quad w \quad \cdots]$. Similarly, the linear programming problem "Minimize CX subject to the constraints $AX \geq B, X \geq \mathbf{0}$" has as its dual "Maximize UB subject to the constraints $UA \leq C, U \geq \mathbf{0}$."

PRACTICE PROBLEMS 4

A linear programming problem involving three variables and four nontrivial inequalities has the number 52 as the maximum value of its objective function.

1. How many variables and nontrivial inequalities will the dual problem have?
2. What is the optimum value for the objective function of the dual problem?

EXERCISES 4

In Exercises 1–6, determine the dual problem of the given linear programming problem.

1. Maximize $4x + 2y$ subject to the constraints

$$\begin{cases} 5x + y \leq 80 \\ 3x + 2y \leq 76 \\ x \geq 0, \quad y \geq 0. \end{cases}$$

2. Minimize $30x + 60y + 50z$ subject to the constraints

$$\begin{cases} 5x + 3y + z \geq 2 \\ x + 2y + z \geq 3 \\ x \geq 0, \quad y \geq 0, \quad z \geq 0. \end{cases}$$

3. Minimize $10x + 12y$ subject to the constraints

$$\begin{cases} x + 2y \geq 1 \\ -x + y \geq 2 \\ 2x + 3y \geq 1 \\ x \geq 0, \quad y \geq 0. \end{cases}$$

4. Maximize $80x + 70y + 120z$ subject to the constraints

$$\begin{cases} 6x + 3y + 8z \leq 768 \\ x + y + 2z \leq 144 \\ 2x + 5y \leq 216 \\ x \geq 0, \quad y \geq 0, \quad z \geq 0. \end{cases}$$

5. Minimize $3x + 5y + z$ subject to the constraints

$$\begin{cases} 2x - 4y - 6z \leq 7 \\ y \geq 10 - 8x - 9z \\ x \geq 0, \quad y \geq 0, \quad z \geq 0. \end{cases}$$

6. Maximize $2x - 3y + 4z - 5w$ subject to the constraints

$$\begin{cases} x + y + z + w - 6 \le 10 \\ 7x + 9y - 4z - 3w \ge 5 \\ x \ge 0, \quad y \ge 0, \quad z \ge 0, \quad w \ge 0. \end{cases}$$

7. The final simplex tableau for the linear programming problem of Exercise 1 appears below. Give the solution to the problem and to its dual.

$$\begin{array}{c} \\ x \\ y \\ M \end{array} \begin{array}{c} \begin{array}{ccccc|c} x & y & u & v & M & \end{array} \\ \left[\begin{array}{ccccc|c} 1 & 0 & \frac{2}{7} & -\frac{1}{7} & 0 & 12 \\ 0 & 1 & -\frac{3}{7} & \frac{5}{7} & 0 & 20 \\ \hline 0 & 0 & \frac{2}{7} & \frac{6}{7} & 1 & 88 \end{array}\right] \end{array}$$

8. The final simplex tableau for the *dual* of the linear programming problem of Exercise 2 appears below. Give the solution to the problem and to its dual.

$$\begin{array}{c} \\ v \\ y \\ z \\ M \end{array} \begin{array}{c} \begin{array}{cccccc|c} u & v & x & y & z & M & \end{array} \\ \left[\begin{array}{cccccc|c} 5 & 1 & 1 & 0 & 0 & 0 & 30 \\ -7 & 0 & -2 & 1 & 0 & 0 & 0 \\ -4 & 0 & -1 & 0 & 1 & 0 & 20 \\ \hline 13 & 0 & 0 & 0 & 0 & 1 & 90 \end{array}\right] \end{array}$$

9. The final simplex tableau for the *dual* of the linear programming problem of Exercise 3 appears below. Give the solution to the problem and to its dual.

$$\begin{array}{c} \\ x \\ v \\ M \end{array} \begin{array}{c} \begin{array}{cccccc|c} u & v & w & x & y & M & \end{array} \\ \left[\begin{array}{cccccc|c} 3 & 0 & 5 & 1 & 1 & 0 & 22 \\ 2 & 1 & 3 & 0 & 1 & 0 & 12 \\ \hline 3 & 0 & 5 & 0 & 2 & 1 & 24 \end{array}\right] \end{array}$$

10. The final simplex tableau for the linear programming problem of Exercise 4 appears below. Give the solution to the problem and to its dual.

$$\begin{array}{c} \\ x \\ z \\ y \\ M \end{array} \begin{array}{c} \begin{array}{ccccccc|c} x & y & z & u & v & w & M & \end{array} \\ \left[\begin{array}{ccccccc|c} 1 & 0 & 0 & \frac{5}{12} & -\frac{5}{3} & \frac{1}{12} & 0 & 98 \\ 0 & 0 & 1 & -\frac{1}{8} & 1 & -\frac{1}{8} & 0 & 21 \\ 0 & 1 & 0 & -\frac{1}{6} & \frac{2}{3} & \frac{1}{6} & 0 & 4 \\ \hline 0 & 0 & 0 & \frac{20}{3} & \frac{100}{3} & \frac{10}{3} & 1 & 10{,}640 \end{array}\right] \end{array}$$

In Exercises 11–14, determine the dual problem. Solve either the original problem or its dual by the simplex method and then give the solutions to both.

11. Minimize $3x + y$ subject to the constraints

$$\begin{cases} x + y \ge 3 \\ 2x \ge 5 \\ x \ge 0, \quad y \ge 0. \end{cases}$$

12. Minimize $3x + 5y + z$ subject to the constraints

$$\begin{cases} x + y + z \ge 20 \\ y + 2z \ge 0 \\ x \ge 0, \quad y \ge 0, \quad z \ge 0. \end{cases}$$

13. Maximize $10x + 12y + 10z$ subject to the constraints

$$\begin{cases} x - 2y \qquad \le 6 \\ 3x \qquad + z \le 9 \\ \qquad y + 3z \le 12 \\ x \ge 0, \quad y \ge 0, \quad z \ge 0. \end{cases}$$

14. Maximize $x + 3y$ subject to the constraints

$$\begin{cases} x + y \le 7 \\ x + 2y \le 10 \\ x \ge 0, \quad y \ge 0. \end{cases}$$

Exercises 15–18 refer to the furniture manufacturing problem. Some of the data from the problem are:

[carpentry]	$6x + 3y \le 96$	Solution: Maximum profit = \$1400
[finishing]	$x + y \le 18$	Solution to dual problem: $u = \frac{10}{3}$, $v = 60$, $w = 0$
[upholstery]	$2x + 6y \le 72$	

15. What is the maximum profit possible if 99 man-hours of labor are available for carpentry?

16. What is the maximum profit possible if the number of man-hours of labor available for finishing is decreased by 1?

17. What is the maximum profit possible if 18.5 man-hours of labor are available for finishing?

18. What is the maximum profit possible if 74 man-hours of labor are available for upholstery?

Exercises 19–22 refer to the nutrition problem (Example 2) from Section 2 of the previous chapter. Some of the data from this problem are:

[protein]	$15x + 22.5y \ge 90$	Solution: Minimum cost is 66 cents
[calories]	$810x + 270y \ge 1620$	Solution to dual problem: $u = \frac{2}{5}$, $v = \frac{1}{54}$, $w = 0$
[riboflavin]	$\frac{1}{9}x + \frac{1}{3}y \ge 1$	

19. What would be the minimum cost possible if only 80 grams of protein were required?

20. How much could be saved by decreasing the calorie requirement to 1512?

21. How much could be saved by decreasing the riboflavin requirement by $\frac{1}{4}$ milligram?

22. What would be the minimum cost possible if 95 grams of protein were required?

23. Give an economic interpretation to the dual of the shipping problem of Exercise 5 of Section 3 of the preceding chapter.

24. Give an economic interpretation to the dual of the mining problem in Exercise 6 of Section 3 of the preceding chapter.

25–30. For each of the linear programming problems in Exercises 1–6, identify the matrices A, B, C, X and state the problem in terms of matrices. Then identify the matrix U and express the dual problem in terms of matrices.

SOLUTIONS TO PRACTICE PROBLEMS 4

1. Four variables and three nontrivial inequalities. The number of variables in the dual problem is always the same as the number of nontrivial inequalities in the original problem. The number of nontrivial inequalities in the dual problem is the same as the number of variables in the original problem.

2. Minimum value of 52. The original problem and the dual problem always have the same optimum values. However, if this value is a maximum for one of the problems, it will be a minimum for the other.

Chapter 7: CHECKLIST

- □ Standard form of linear programming problem
- □ Slack variable
- □ Group I, group II variables
- □ Simplex tableau
- □ Simplex method for problems in standard form
- □ Converting minimization problems to maximization problems
- □ Reduction of linear programming problems to standard form
- □ Dual problem
- □ Fundamental theorem of duality
- □ Matrix formulation of linear programming problem

Chapter 7: SUPPLEMENTARY EXERCISES

Use the simplex method to solve the following linear programming problems.

1. Maximize $3x + 4y$ subject to the constraints

$$\begin{cases} 2x + y \leq 7 \\ -x + y \leq 1 \\ x \geq 0, \quad y \geq 0. \end{cases}$$

2. Maximize $2x + 5y$ subject to the constraints

$$\begin{cases} x + y \leq 7 \\ 4x + 3y \leq 24 \\ x \geq 0, \quad y \geq 0. \end{cases}$$

3. Maximize $2x + 3y$ subject to the constraints

$$\begin{cases} x + 2y \leq 14 \\ x + y \leq 9 \\ 3x + 2y \leq 24 \\ x \geq 0, \quad y \geq 0. \end{cases}$$

4. Maximize $3x + 7y$ subject to the constraints

$$\begin{cases} x + 2y \le 10 \\ 4x + 3y \le 30 \\ -2x + y \le 0 \\ x \ge 0, \quad y \ge 0. \end{cases}$$

5. Minimize $x + y$ subject to the constraints

$$\begin{cases} 7x + 5y \ge 40 \\ x + 4y \ge 9 \\ x \ge 0, \quad y \ge 0. \end{cases}$$

6. Minimize $3x + 2y$ subject to the constraints

$$\begin{cases} x + y \ge 6 \\ x + 2y \ge 0 \\ x \ge 0, \quad y \ge 0. \end{cases}$$

7. Minimize $20x + 30y$ subject to the constraints

$$\begin{cases} x + 4y \ge 8 \\ x + y \ge 5 \\ 2x + y \ge 7 \\ x \ge 0, \quad y \ge 0. \end{cases}$$

8. Minimize $5x + 7y$ subject to the constraints

$$\begin{cases} 2x + y \ge 10 \\ 3x + 2y \ge 18 \\ x + 2y \ge 10 \\ x \ge 0, \quad y \ge 0. \end{cases}$$

9. Maximize $36x + 48y + 70z$ subject to the constraints

$$\begin{cases} x \le 4 \\ y \le 6 \\ z \le 8 \\ 4x + 3y + 2z \le 38 \\ x \ge 0, \quad y \ge 0, \quad z \ge 0. \end{cases}$$

10. Maximize $3x + 4y + 5z + 4w$ subject to the constraints

$$\begin{cases} 6x + 9y + 12z + 15w \le 672 \\ x - y + 2z + 2w \le 92 \\ 5x + 10y - 5z + 4w \le 280 \\ x \ge 0, \quad y \ge 0, \quad z \ge 0, \quad w \ge 0. \end{cases}$$

11. Determine the dual problem of the linear programming problem in Exercise 3.

12. Determine the dual problem of the linear programming problem in Exercise 7.

13. The final simplex tableau for the linear programming problem of Exercise 3 appears below. Give the solution to the problem and to its dual.

$$\begin{array}{c} \\ y \\ x \\ w \\ M \end{array} \begin{array}{c} \begin{array}{cccccc|c} x & y & u & v & w & M & \end{array} \\ \left[\begin{array}{cccccc|c} 0 & 1 & 1 & -1 & 0 & 0 & 5 \\ 1 & 0 & -1 & 2 & 0 & 0 & 4 \\ 0 & 0 & 1 & -4 & 1 & 0 & 2 \\ \hline 0 & 0 & 1 & 1 & 0 & 1 & 23 \end{array}\right] \end{array}$$

14. The final simplex tableau for the *dual* of the linear programming problem of Exercise 7 appears below. Give the solution to the problem and to its dual.

$$\begin{array}{c} \\ v \\ u \\ \\ \end{array} \begin{array}{c} \begin{array}{cccccc|c} u & v & w & x & y & M & \end{array} \\ \left[\begin{array}{cccccc|c} 0 & 1 & \frac{7}{3} & \frac{4}{3} & -\frac{1}{3} & 0 & \frac{50}{3} \\ 1 & 0 & -\frac{1}{3} & -\frac{1}{3} & \frac{1}{3} & 0 & \frac{10}{3} \\ \hline 0 & 0 & 2 & 4 & 1 & 1 & 110 \end{array}\right] \end{array}$$

15, 16. For each of the linear programming problems in Exercises 3 and 7, identify the matrices A, B, C, X and state the problem in terms of matrices. Then identify the matrix U and express the dual problem in terms of matrices.